Philipp Hansmann

LDA+DMFT: From bulk to heterostructures

Philipp Hansmann

LDA+DMFT: From bulk to heterostructures

realistic calculations and novel methods for the treatment of electronic correlation in solids

Südwestdeutscher Verlag für Hochschulschriften

Impressum/Imprint (nur für Deutschland/ only for Germany)
Bibliografische Information der Deutschen Nationalbibliothek: Die Deutsche Nationalbibliothek verzeichnet diese Publikation in der Deutschen Nationalbibliografie; detaillierte bibliografische Daten sind im Internet über http://dnb.d-nb.de abrufbar.

Alle in diesem Buch genannten Marken und Produktnamen unterliegen warenzeichen-, marken- oder patentrechtlichem Schutz bzw. sind Warenzeichen oder eingetragene Warenzeichen der jeweiligen Inhaber. Die Wiedergabe von Marken, Produktnamen, Gebrauchsnamen, Handelsnamen, Warenbezeichnungen u.s.w. in diesem Werk berechtigt auch ohne besondere Kennzeichnung nicht zu der Annahme, dass solche Namen im Sinne der Warenzeichen- und Markenschutzgesetzgebung als frei zu betrachten wären und daher von jedermann benutzt werden dürften.

Verlag: Südwestdeutscher Verlag für Hochschulschriften Aktiengesellschaft & Co. KG
Dudweiler Landstr. 99, 66123 Saarbrücken, Deutschland
Telefon +49 681 37 20 271-1, Telefax +49 681 37 20 271-0
Email: info@svh-verlag.de
Zugl.: Wien, Technische Universität Wien, Diss., 2010

Herstellung in Deutschland:
Schaltungsdienst Lange o.H.G., Berlin
Books on Demand GmbH, Norderstedt
Reha GmbH, Saarbrücken
Amazon Distribution GmbH, Leipzig
ISBN: 978-3-8381-1839-0

Imprint (only for USA, GB)
Bibliographic information published by the Deutsche Nationalbibliothek: The Deutsche Nationalbibliothek lists this publication in the Deutsche Nationalbibliografie; detailed bibliographic data are available in the Internet at http://dnb.d-nb.de.

Any brand names and product names mentioned in this book are subject to trademark, brand or patent protection and are trademarks or registered trademarks of their respective holders. The use of brand names, product names, common names, trade names, product descriptions etc. even without a particular marking in this works is in no way to be construed to mean that such names may be regarded as unrestricted in respect of trademark and brand protection legislation and could thus be used by anyone.

Publisher: Südwestdeutscher Verlag für Hochschulschriften Aktiengesellschaft & Co. KG
Dudweiler Landstr. 99, 66123 Saarbrücken, Germany
Phone +49 681 37 20 271-1, Fax +49 681 37 20 271-0
Email: info@svh-verlag.de

Printed in the U.S.A.
Printed in the U.K. by (see last page)
ISBN: 978-3-8381-1839-0

Copyright © 2010 by the author and Südwestdeutscher Verlag für Hochschulschriften Aktiengesellschaft & Co. KG and licensors
All rights reserved. Saarbrücken 2010

Für meine Eltern
Regina & Peter

Contents

1 **Introduction & Scope**   1

2 **Combining DFT and many-body approaches**   7
  2.1 Density Functional Theory and its Local Density Approximation   11
  2.2 Dynamical Mean Field Theory 15
  2.3 LDA+DMFT 23
    2.3.1 Obtaining the local self energy on the real axis 28
  2.4 Observables from LDA+DMFT 30
    2.4.1 Angular Resolved Photoemission Spectroscopy (ARPES) . 31
    2.4.2 Optical Conductivity (IR) 33
    2.4.3 X-ray Absorption Spectroscopy (XAS) 40

3 **Bulk 3d-transition metal compounds: Theory vs. Experiment**   43
  3.1 Optics and X-ray absorption of V_2O_3 44
    3.1.1 The story so far 45
    3.1.2 The low energy t_{2g} NMTO model 51
    3.1.3 Optical conductivity: Phase separation around the MIT . 53
    3.1.4 X-ray absorption on the V-K-edge: Pressure vs. doping . 65
  3.2 Optics of $NiSe_xS_{1-x}$ 83

4 **Nickel oxide superstructures**   95
  4.1 Starting from the high T_C cuprates 96
  4.2 $LnSrNiO_4$ perovskites 104
  4.3 $LaAlO_3/LaNiO_3$ 115
    4.3.1 MIT in a quarter-filled two-band system 123

Contents

5 Expanding the basis set   **139**
    5.1   LDA+HartreeDMFT 140
          5.1.1   Implementation of HartreeDMFT for the nickelates 141
          5.1.2   The cuprate "(extended) Emery" model: a revision 157
    5.2   Multilayer DMFT 167
          5.2.1   periodic stacks 172

6 Summary & Outlook   **175**

1 Introduction & Scope

Correlations of various types are present in everyday life: Stock quotations, traffic jams, or a soccer match are examples for correlated systems. In a mathematical way we can define the concept of correlation as the causal relation between two (comparable) entities.
In the case of solid state physics, such entities could be for example electronic densities and we refer to an electron system as being correlated when

$$\langle \hat{\rho}(\mathbf{r})\hat{\rho}(\mathbf{r}')\rangle \neq \langle \hat{\rho}(\mathbf{r})\rangle\langle \hat{\rho}(\mathbf{r}')\rangle$$

where $\hat{\rho}(\mathbf{r})$ is the electron density operator and the outer brackets express the expectation value of the operator. The inequality means, that the expectation value of the density at a certain coordinate r is not independent from the density at another coordinate r' – they are correlated.
This thesis is devoted to current forefront research in the field of theoretical modeling for correlated solids. The correlated compounds have become the focus of an ever-growing community in the last decades. The reason for the great popularity of such systems can be found, on the one hand, in the highly bundled variety of fascinating physics, which manifests itself in rich phase diagrams and highly non-trivial ground states. On the other hand the correlated systems bear interesting technological applicability which is concomitant with their sensitivity to external parameters. The unusual sensitivity to small perturbations, such as the addition/subtraction of a few electrons, the application of external fields, etc., is also reflected in the terminology of the observed effects which is packed with intensifying adjectives: large

1 Introduction & Scope

resistivity changes, huge volume changes i.e. collapses, high T_C superconductivity, strong thermoelectric response, colossal magnetoresistance, and gigantic non–linear optical effects are examples for the huge susceptibilities, i.e. response functions, of the correlated materials.

Since John Hubbard introduced his famous Hamiltonian for the description of correlation effects in d– and f–bands in 1963 [87] we have witnessed an amazing development, theoretically as well as experimentally. Modern spectroscopy techniques like angular resolved photoemission spectroscopy, measurements of the optical conductivity or more advanced ellipsometry, x–ray absorption spectroscopy, neutron scattering experiments, various microscopy techniques, and many more related approaches or combinations of them yield data which have to be interpreted and understood by means of reliable theoretical calculations. To provide such theoretical tools for realistic calculations is, however, an extremely challenging task. Correlated materials often exhibit non–Fermi liquid behavior and many–body effects are not negligible anymore. In that respect, the *merger* of the density functional theory (DFT) within its local density approximation (LDA), probably the most successful technique in materials calculations of the last century, and the dynamical mean field theory (DMFT), a non–perturbative many–body quantum field approach, has turned out to be extremely successful and became popular as LDA+DMFT approach.

Moreover, a constant improvement of understanding and the increasing knowledge about correlated systems gives us the possibility to engineer materials according to our needs – thanks to the progress in synthezising technology the term *material design* developed from a catchphrase to reality. In turn, this progress also calls for the constant development and extension of theoretical methods in order to capture the vast amount of new experimental data.

In the second chapter of this thesis we will give a pedagogical overview of the methods that were employed starting with a brief sketch of density functional theory and LDA. In the following, we will derive the DMFT equations in a diagrammatic way and explain the self consistent mapping on auxiliary Anderson impurity models. Finally we discuss the combination of both approaches to a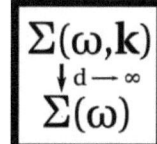
"state of the art" LDA+DMFT scheme. As a last section of the second chapter we give a brief summary of experimentally accessible observables which we can also calculate within the LDA+DMFT scheme. This last section is intended as a reference for the experiment – theory comparisons we will encounter in the following chapters.

Chapter three is devoted to the metal–to–insulator transition (MIT) of strongly correlated systems. The insulating phase caused by electronic correlation is categorized as a Mott Insulator. Close to the transition point the metallic state often shows fluctuations and sometimes orderings in the spin, charge, and orbital degrees of freedom. The properties of these metals are frequently quite dif-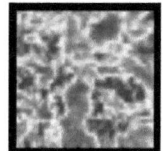
ferent from those of ordinary metals, as measured by transport, optical, and magnetic probes. The DMFT approach turned out to be extremely well suited in order to describe the Mott metal–to–insulator transition. Specifically, we will provide new insight in the MITs of two very well known correlated compounds: Vanadium sesquioxide V_2O_3 and nickel disulfide NiS_2 which are, in the undoped form a correlated metal and a charge transfer insulator, respectively.

V_2O_3 undergoes a transition to a Mott insulator upon doping with Cr. A combined analysis of experimental data for the optical conductivity and LDA+DMFT results is employed to clarify details of the MIT. We will show experimental and theoretical evidence that the metallic phase in the vicinity of the MIT cannot be described by a homogeneous phase but has to be accounted for as a mixed phase state. Furthermore, the experimental measurements have shown that the long established "common wisdom" that for V_2O_3 the doping effect could be reversed with the application of an external

1 Introduction & Scope

pressure has to be abandoned. The metallic state of the undoped compound and that of the Cr doped compound under pressure display very different spectra. In the following theoretical analysis we also quantify the difference of the underlying ground states with the help of vanadium K–edge hard x–ray absorption spectra: For the interpretation of the spectrum we introduce a novel combination of LDA+DMFT calculations with parameter–based full–multiplet configuration interaction calculations. This new approach allows us to exploit excitonic features in the pre–edge region of the vanadium K–edge as a ground state probe even for the system under pressure – this was not possible before, since the soft x–rays of vanadium L–edge measurements (previously used as a ground state probe) are strongly absorbed by the diamond pressure cells.

The second transition we discuss is the insulator to metal transition of the charge transfer insulator nickel disulfide NiS_2 upon Selenium doping or application of external pressure. Similar to the case of V_2O_3, the idea of a doping/pressure equivalence was suggested. Yet, also in the case of NiS_2, careful analysis shows that this concept is not consistent with experimental and theoretical results. In particular we will discuss results of band structure calculations and show that the different microscopic mechanisms which drive the transition upon doping or pressure application can be understood to some extent on the LDA level.

The focus of chapter four is the possibility of finding cuprate-analogous high temperature superconductivity in novel nickel-based bulk- or heterostructures. While in the third chapter we presented new physics of known systems, in chapter four we try to exploit known mechanisms of the superconducting cuprates in order to suggest new systems as candidates for high T_C superconductivity.

In the first section of the chapter we review results by Pavarini *et al.* [154]. In this work the authors describe the empirical observation that the critical temperature of cuprate superconductors is correlated to the interplay of the *planar* (Cu $x^2 - y^2$) conduction band around the Fermi energy ε_F and an *axial* (mainly Cu $4s$) band at higher energies. In the following sections we discuss the potential of nickel based compounds to display a similar *planar* band *axial* band scenario.

We will start with the "bulk" LnSrNiO$_4$ (Ln=La, Nd, Eu) series. Band structure calculations for these systems show indeed the presence of a *planar* and an *axial* band. However, both bands are residing at ε_F. The basic idea is, that correlation effects will push the *axial* band up in energy, thus leaving the *planar* band alone around the Fermi energy which is precisely the situation realized in the cuprates. Our LDA+DMFT study shows that the correlation effect indeed enhances the splitting of the bands, but in the wrong direction. We remain with the *axial* band at ε_F. Yet, we also observe that the LnSrNiO$_4$ systems are somewhat on the edge, and it depends sensitively on the details of the structure which of the two bands is pushed up by correlation effects.

This leads us to nickel based heterostructures which are synthezied with the help of cutting–edge technology. In these compounds the structural parameters are more controllable from the "outside" and details of the synthesis procedure, like the choice of substrates or chemical substitution, allow for an actual *material design*. Specifically with our LDA+DMFT calculations we will focus on the 1/1 layered system LaAlO$_3$/LaNiO$_3$ for which we find the desired *planar* band to remain at the Fermi energy generating a cuprate like Fermi surface structure. Together with antiferromagnetic fluctuations, which we observe in a small toy model for the system at low temperatures, we conclude that with these features, the nickel based heterostructures incorporate the basic ingredients for high T_C superconductivity.

The last chapter of this thesis focuses on methodological advances which were developed in the context of this work. Our extensions of the standard DMFT are based on the expansion of the DMFT basis set. The first extension we discuss is the so called HartreeDMFT. Sometimes the correlated states around the Fermi energy are entangled with less correlated ones – we find a typical 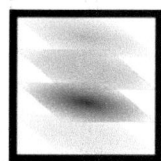 example in transition metal oxides, where correlated 3d states strongly hybridize with the Oxygen 2p degrees of freedom. Often this hybridization can be taken into account by analyzing *effective* d–bands which incorporate the

1 Introduction & Scope

hybridization effects implicitly. However, in some cases this hybridization is too strong and the p–states are too important to be taken into account by effective d–state models. Charge–transfer insulators are also a good example where d–only models cannot capture the physics correctly, since their spectral gap is formed between d– and p–states. In the HartreeDMFT approach we divide the full Hamiltonian into a subspace for which we perform the self consistent DMFT scheme (i.e. for the d–states) and another subspace, where less correlated, but nonetheless important, states (i.e. the p–states) are treated with a less expensive Hartree mean field scheme. As a pedagogical introduction we will discuss a dp–model for the nickel based heterostructures from chapter four. Next we will turn to the parent compound of the high T_C cuprates $LaCuO_4$ which is a charge transfer insulator. In this context we will discuss the LDA+HartreeDMFT analysis of Emery–like three band models and elaborate on inconsistencies of some assumptions done in recent studies.

The second extension we will discuss is connected to heterostructure systems: A new DMFT self consistent scheme for multilayer systems. Also this step can be understood as an extension of the DMFT basis set. This time, in order to capture features of heterostructures which are beyond the standard DMFT scheme. The multilayer scheme can be seen as the simultaneous computation of several DMFT problems, one for each subsystem, i.e. for each layer, while the subsystems hybridize through the self consistent iterations of the layerDMFT scheme. Besides discussing the implementation pedagogically, we show the result for a periodic ...ABABA... system of two insulators (a Mott and a band insulator) which, upon bringing them together, start to show metallic behavior.

Throughout the text the reader will find some boxes labeled with "**Info**". These boxes contain details of technical information not necessary for the understanding of the main discussion.

2 Combining DFT and many-body approaches

For condensed matter physics it is in principle easy to write down the electronic Hamiltonian for the 'Theory of Everything':

$$\hat{H} = \sum_\sigma \int d^3r \hat{\Psi}^\dagger(\underline{r},\sigma) \left[-\frac{\hbar^2 \Delta_i}{2m_e} + \sum_l \frac{-e^2}{4\pi\varepsilon_0} \frac{Z_l}{|\underline{r}_i - \underline{R}_l|} \right] \hat{\Psi}(\underline{r},\sigma) +$$

$$\frac{1}{2} \sum_{\sigma,\sigma'} \int d^3r d^3r' \hat{\Psi}^\dagger(\underline{r},\sigma) \hat{\Psi}^\dagger(\underline{r}',\sigma') \frac{-e^2}{4\pi\varepsilon_0} \frac{1}{|\underline{r}_i - \underline{r}_j|} \hat{\Psi}(\underline{r}',\sigma') \hat{\Psi}(\underline{r},\sigma) \quad (2.1)$$

where $\underline{r}_i$ and $\underline{R}_l$ denote the position of electron i and ion l with charge $-e$ and $Z_l e$ respectively; Δ_i is the Laplace operator for the kinetic energy of electrons with mass m_e; ε_0 and $\hbar$ are the vacuum dielectric and Planck constant; further relativistic corrections were neglected.

However, even a numerical solution of this full Hamiltonian is far from being possible if the particles considered exceed a very small number also after the electronic degrees of freedom have been decoupled from the lattice part of the Hamiltonian with the so called Born–Oppenheimer approximation [12]. The main problems arise from the two-particle-operator for the Coulomb interaction between the electrons involved (last term in Hamiltonian (2.1)).
The starting point for understanding the electronic structure and excitation spectra of solids in a standard textbook is usually the treatment of noninter-

2 Combining DFT and many-body approaches

acting independent (i.e. uncorrelated) electrons. At a first glance this may seem completely unjustified, since the energy scale for their Coulomb interaction is of the order of eVs and thus far from being negligible compared to the other energy scales. It is only the Landau Fermi Liquid theory [1], which explains the very remarkable fact that in many cases the tough problem of correlated particles can be mapped onto an independent particle problem with effective single particle potentials. Hence, it must always be kept in mind, that these single particle potentials are not simply the ones from the ionic background but, in fact, *effective* potentials which are supposed to include the correlation part of the problem. In this way also the particles must no longer be viewed as independent but as "effective" or "particle–like" entities: the *quasiparticles*.

Within the so called Density Functional Theory (DFT) [86] and its Local Density (LDA) or related approximations, the mapping onto effective single particle potentials has proven to be extremely successful for a huge variety of compounds. As well as this approach works for many compounds, if considerable correlations are at present - like in the transition metals or rare–earths and their oxides or actinides – LDA may yield wrong results. As long as the Fermi liquid picture remains valid, a renormalization of LDA results can still give reasonable insight to the excitations of the system. If, however, at a certain point the entire concept of "effective" single particle states of the Fermi liquid theory breaks down the LDA is not applicable anymore. Examples for such systems are the vanadium oxides VO_2 and V_2O_3, NiO, and the parent compound of the high T_C cuprates La_2CuO_4 which are predicted to be metallic above their Neél temperature by LDA when they are, in fact, Mott insulators [112, 219, 156, 121, 121]. But not only oxides are challenging compounds as we will see in section 3.2 where we discuss the interesting case of the MIT in nickel disulfide NiS_2.

From the DFT community one of the first successful attempts to expand the LDA formalism in order to treat strong electronic correlations *ab initio* was the introduction of the so called LDA+U method by Anisimov *et al.* [9] where an orbital dependent potential, calculated in the Hartree approximation, is employed to include the correlation effects.

The LDA+U approach will lead to insulating spectra in the presence of orbital or magnetic ordering which look similar to those observed for a paramagnetic Mott–insulator. However, the insulating spectra of LDA+U have some delicate shortcomings: For a paramagnetic Mott–insulator the LDA+U solution has a too low entropy since it has to be a symmetry broken ordered solution. But even for an ordered system the LDA+U spectra are not accurate, since the excited states (e.g. spin polarons) [19, 20, 211, 120, 24] and the resulting additional spectral features are not captured [181]. Further, strongly correlated metallic phases, which are realized in many transition metal compounds (e.g. also via doping or appliance of pressure) are also beyond the LDA+U scheme since its solutions almost automatically yields localized electrons with a gapped spectrum.

The problem of strongly correlated systems was, on the other hand, attacked from a different side: the *many body model Hamiltonian* community. Instead of approximately solving Hamiltonian (2.1) like in DFT, a simpler, parameter–based model Hamiltonian is constructed and solved with the aim of understanding the underlying physical mechanisms of the systems characteristics. The results of these calculations, however, are often far from a fully realistic description and their parameters have to be obtained e.g. by a fit to experimental results. One of the most prominent starting points for the model Hamiltonians is the so called Hubbard model. Derived from the continuum formulation (2.1) and translated to the basis of e.g. Wannier functions on a lattice with the assumption that the Coulomb interaction is purely local we can formulate it like:

$$\hat{H} = \sum_{iljm\sigma} t_{il,jm} \hat{c}^\dagger_{il\sigma} \hat{c}_{jm\sigma} + \sum_{ilmno\sigma\sigma'} U_{lmno} \hat{c}^\dagger_{il\sigma} \hat{c}^\dagger_{im\sigma'} \hat{c}_{in\sigma'} \hat{c}_{io\sigma} \qquad (2.2)$$

here $\hat{c}^\dagger_{il\sigma}(\hat{c}_{il\sigma})$ creates (annihilates) an electron with spin σ and orbital index l at lattice site i; $t_{il,jm}$ is a hopping amplitude between lattice sites i and j and (Wannier–)orbitals l and m; finally, U_{lmno} denotes a general local Coulomb interaction. Let us remark that the original Hubbard model [87] is a single band model, whereas Eq.(2.2) is already a multiband extension and often referred to as multiband–Hubbard– or Kanamori Hamiltonian. However, throughout

2 Combining DFT and many-body approaches

this work, the reference to a "Hubbard Hamiltonian" is meant to include the multiband extensions unless specifically stated otherwise. Actually, Hamiltonian (2.2) gives a very intuitive picture, since the two relevant energy scales – that is the kinetic ($t_{il,jm}$) and the Coulomb (U_{lmno}) part – are included explicitely: Electrons hop around on the lattice and each time they occupy the same site it costs a "Coulomb" energy U. However, it is easy to realize that also this Hamiltonian is far from being easy solvable, (apart from the one dimensional case with the Bethe Ansatz) since its kinetic part is diagonal in momentum space, whereas the interaction part is diagonal in real space. If not one of the energy scales is dominating the other – like in the case of the correlated materials – there is no obvious small paramter for a perturbation expansion or other simplifications: the two terms have to be treated on equal footing. Nonetheless, it is also easy to guess that a solution of such a case will be rewarded by rich phasediagrams and fascinating physics.

The method which has proven to be capable of solving Hamiltonians like (2.2) within a controlled, and non–perturbative, approximation is the dynamical mean–field theory (DMFT) [128, 136]. The DMFT can be seen as the quantum extension of a classical mean field theory and, just like its classical counterpart, it becomes exact in the limit of infinite dimensions (infinite coordination number) where the self energy only depends on the frequency. The DMFT is "dynamic" as opposed to "static" (Hartree Fock type) mean field theories with a constant, i.e. not time/frequency dependent, self energy. The success of the DMFT is also due to the possibility, in the limit of inifinite dimensions, of mapping the Hubbard model, and other many body models, self consistently on an auxiliary Anderson impurity model (AIM) as it was shown by Georges and Kotliar [67]. This mapping allows for the use of well established AIM solvers within the DMFT calculation.

On the one hand there is the LDA approach from the DFT community, which will fail for the strongly correlated systems, and on the other hand, from the many body community, the DMFT solver of parameter based (i.e. non– *ab initio*) Hamiltonians like (2.2). Hence, there is a very natural motivation to combine the strenghts of both methods and join forces. In the following sections first the LDA, then the DMFT, and finally their combination to

2.1 Density Functional Theory and its Local Density Approximation

LDA+DMFT are described seperately in more detail.

2.1 Density Functional Theory and its Local Density Approximation

As long as only **ground state properties** *are considered there exists an effective one–particle potential* $V^{\text{eff.}}[\rho(\underline{r})]$, *which is a functional of the ground state density, such that the complete many body problem (including also the non–mean–field–like interactions) becomes an effective single particle problem solvable by single particle Schrödinger equations.* This claim (for the ground-state properties) is exact and known as the Hohenberg–Kohn theorem [86]. Moreover, the theorem states that for any observable $\mathcal{O}$, the ground state expectation value $\langle\mathcal{O}\rangle_{\text{gs}} = \mathcal{F}_\mathcal{O}[\rho_{\text{gs}}]$ is a functional of the ground state density only, and that there exists a Ritz–variational scheme for the ground state energy, i.e., $E[\rho_{\text{gs}}] < E[\rho]$ for all ρ.

However, in physics, the complexity of a problem is often a conserved quantity and one soon realizes that the Hohenberg–Kohn theorem is not the final solution of everything. Although the theorem proofs its existence, the effective potential we seek is unknown. Yet, the theorem opens up the following strategy: First of all, separate the functional for the ground state energy into known and unknown parts:

$$E[\rho] = E_{\text{kin.}}[\rho] + E_{\text{ion}}[\rho] + E_{\text{hartree}}[\rho] + E_{\text{xc}}[\rho] \qquad (2.3)$$

where $E_{\text{ion}}[\rho] = \int d^3r V_{\text{ion}}(\underline{r})\rho(\underline{r})$ is the energy of the electrons in the potential of the ions, i.e., the external potential, $E_{\text{hartree}}[\rho] = \int\int d^3r'd^3r V_{\text{ee}}(\underline{r}-\underline{r}')\rho(\underline{r}')\rho(\underline{r})$ is the Hartree mean field part of the inter–electronic Coulomb interaction, and $E_{\text{xc}}[\rho]$ represents the unknown quantity of the DFT which remains[1]. Now,

---
[1] An interesting fact to remark here is, that the quantity $E[\rho]-E_{\text{ion}}[\rho]$ is material independent (see [98] for a proof), so that, if we would know the DFT functional for one material we could calculate all materials by simply adding $E_{\text{ion}}[\rho]$.

2 Combining DFT and many-body approaches

in principle, the exchange term could be approximated, and the variational scheme for the ground state could be performed. But one would still have to face the problem of expressing the kinetic energy $E_{\text{kin.}}$ through the density $\rho(\underline{r})$. In order to avoid this complication Kohn and Sham [86] suggested to implement a fictitious non–interacting system, whose external potential is determined in such a way that the real and the fictitious systems have the same density. This density is then:

$$\rho(\underline{r}) = \sum_{i=1}^{N} |\phi_i(\underline{r})|^2 \qquad (2.4)$$

where the ϕ_i represent a set of *auxiliary* one-particle wave functions. Then the variational minimization is performed with respect to the ϕ_i instead of $\rho(\underline{r})$. Hence, we minimize $\delta\{E[\rho] - \varepsilon_i[\int d^3r|\phi_i(\underline{r})|^2] - 1\}/\delta\phi_i(\underline{r})$ where the Lagrange parameters ε_i take care of the normalization of the ϕ_i. This minimization leads us to the well known Kohn–Sham equations:

$$\left[-\frac{\hbar^2}{2m_e}\Delta + V_{\text{ion}}(\underline{r}) + \int d^3r V_{\text{ee}}(\underline{r} - \underline{r}')\rho(\underline{r}') + \frac{\delta E_{\text{xc}}[\rho]}{\delta\rho(\underline{r})}\right]\phi_i(\underline{r}) = \varepsilon_i\phi_i(\underline{r}) \qquad (2.5)$$

We realize, that these equations are, in fact, single particle Schrödinger equations with a time–averaged potential

$$V_{\text{eff.}} = V_{\text{ion}}(\underline{r}) + \int d^3r V_{\text{ee}}(\underline{r} - \underline{r}')\rho(\underline{r}') + \frac{\delta E_{\text{xc}}[\rho]}{\delta\rho(\underline{r})} \qquad (2.6)$$

These Kohn–Sham equations are still exact since we have not yet made any approximation for the interaction part which is encoded in the exchange enegry functional. That leaves us with the solution of equations (2.5) which is done in a self–consistent way sketched in Fig. 2.1

1. Choose a starting density $\rho(\underline{r})$

2. Calculate the effective single particle potential $V_{\text{eff.}}(\underline{r})$

2.1 Density Functional Theory and its Local Density Approximation

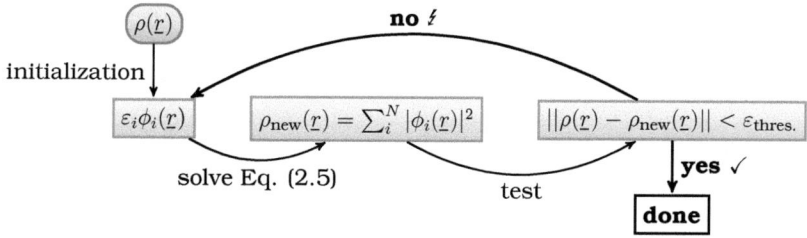

Fig. 2.1: Density functional self consistency cycle

3. Solve the Schrödinger equations (2.5) to obtain the ϕ_i – since in solids the effective potential will be periodic in the lattice we employ the standard band theory methods (see e.g. chapter 4 of [43]) for this step.

4. Evaluate the new density from the ϕ_i with (2.4)

5. Compare old and new density and close the self consistency (see Fig. 2.1)

It should be stressed at this point that, strictly speaking, the single particle wave–functions ϕ_i have no right to be seen as physical quantities. Basically, they represent a reference system of decreased sophistication constructed to retain the ground state density. Neither the spectrum nor the eigenfunctions of the Kohn–Sham equations should have in principle any relation with the corresponding quantities of the true system[2]. Nonetheless these identifications give very often surprisingly good results and one of the main applications of DFT is, in fact, the calculation of band–structures, i.e., the interpretation of the Langrange parameters ε_i in Eq. (2.5) as the actual real excitations of the system.

What is missing, in order to perform an actual calculation, is a reasonable approximation for the exchange–correlation potential. This task in itself is a

---
[2] For example it is easy to see that two–particle operators like the Coulomb interaction in Hamiltonian (2.1) or (2.2) will yield eigenstates that are not expressible as single Slater–determinants and thus outside such a single electron basis–set.

2 Combining DFT and many-body approaches

major subject to ongoing research where, more or less empirical functionals are introduced to capture the specific features of a given system – it is, however, not subject of this thesis.

One of the best known and commonly adapted approximation is the one of the *local density approximation* (LDA):

$$E_{xc}^{LDA} = \int d^3 r \rho(\underline{r}) \varepsilon_{xc}^{LDA}(\rho(\underline{r})) \qquad (2.7)$$

where $\varepsilon_{xc}^{LDA}(\rho(\underline{r}))$ is the exchange–correlation energy density of the homogeneous electron gas (taken as a reference system) evaluated at the same density as the true system under consideration, locally evaluated at the position $\underline{r}$. As a consequence the exchange energy $E_{xc}[\rho(\underline{r})]$ which is generally a functional of the density ρ reduces to (2.7). Thus the LDA Hamiltonian can be written as:

$$\hat{H}_{LDA} = \sum_\sigma \int d^3 r \hat{\Psi}^{\sigma\dagger}(\underline{r}) \left[-\frac{\hbar^2}{2m_e} \Delta + V_{ion}(\underline{r}) + \int d^3 r' V_{ee}(\underline{r} - \underline{r}') \rho(\underline{r}') + \frac{\partial E_{xc}^{LDA}(\rho)}{\partial \rho(\underline{r})} \right] \hat{\Psi}^\sigma(\underline{r}) \qquad (2.8)$$

(note that the functional derivative of the Kohn–Sham equations (2.5) became a partial derivative in the LDA). In practice, $\varepsilon_{xc}^{LDA}(\rho(\underline{r}))$ is calculated from the perturbative solution [79, 210] or the numerical simulation [32] of the jellium model, which is defined by $V_{ion}(\underline{r}) =$ const. and thus will yield a constant density $\rho(\underline{r}) = \rho_0$ (which justifies LDA for the jellium). In real materials, especially for the correlated transition metal- or rare–earth compounds this assumption no longer holds and the LDA is bound to fail.

Today, there exist a large number of DFT implementations. Beside the mentioned alternative exchange–correlation potentials, the actual implementations of a given approach (e.g. LDA) differ in their choice of a particular basis–set. And the choice of the basis–set is, in fact, crucial when keeping in mind the motivation to combine the DFT and the many body approaches:

Correlation effects beyond the LDA involve in particular local Coulomb interactions which we want to describe by means of the Hubbard Hamiltonians (2.2). This means, that we will have to choose a suitable basis to define the interaction parameter U_{lmno} on. The concept of the Wannier functions (see e.g. [12]) offers a valid option for such a basis-set. Generally, either a transformation (projection) of plane–waves to the localized Wannier functions has to be performed, or a DFT implementation which already works itself on a localized basis should be employed like the muffin–tin orbitals (MTO) either in the linear version (LMTO) [2] or the N–th order one (NMTO) [3].

2.2 Dynamical Mean Field Theory

As already mentioned, the conventional approach to treat correlated system expressed in terms of, e.g., a Hubbard model (2.2) is the perturbative expansion around a small parameter. This can be done either starting from the limit of zero Coulomb interaction $U = 0$ (weak coupling) or from the opposite limit of zero kinetic energy $t_{\text{hopping}} = 0$, i.e., bandwidth $W = 0$ (strong coupling). However, the most interesting physics happens in the case when both energy scales, interacting and kinetic, are of the same order of magnitude. Yet, in these cases there is no obvious small parameter. In 1989 Metzner and Vollhardt [128] introduced a new limit to improve the description of correlated electron systems: the limit of infinite dimensions $d \to \infty$ (in a sense, the small parameter here is d^{-1}) or equivalently the limit of infinite coordination number in the lattice. In this limit, the competition between kinetic energy W and Coulomb interaction U is maintained, but the self–energy of the problem becomes entirely local $\Sigma(\mathbf{k}, \omega) \to \Sigma(\omega)$, i.e., momentum–independent. It has then been shown by Müller–Hartmann [136, 137, 138] that only the local Coulomb interaction yields dynamic, i.e., ω–dependent, correlations whereas the non–local density–density interactions are reduced to the non–ω–dependent Hartree contribution. Beside the exact solution of the Falikov–Kimball model [57] and the Kondo lattice model for classical spins, the most

2 Combining DFT and many-body approaches

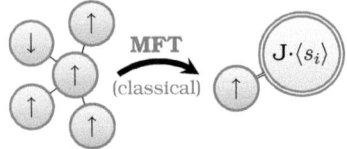

Fig. 2.2: Weiss Mean Field Theory: Instead of coupling to all spins the spin s_i is coupled to a spatially and temporally averaged spin **S**

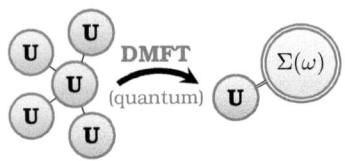

Fig. 2.3: Dynamical Mean Field Theory: The lattice problem is mapped onto a local problem coupled to an effective self energy which still carries the frequency, i.e., temporal information

important step was done in the work by Georges and Kotliar [67] (see also [92, 143, 142]) who showed that a many–body model like the Hubbard model (2.2) could be mapped self consistently on an auxiliary Anderson impurity model (AIM) for the limit of $d \to \infty$. This step was insofar crucial as there are many well established solvers for the AIM which could now be used as solver for the many–body lattice problem within the DMFT approach. Solvers that could be employed are for example iterated perturbation theory [67], self-consistent perturbation theory [138, 184], and the non-crossing approximation [95, 96, 161, 162], numerically exact solvers like quantum Monte Carlo simulations [93, 173, 69], exact diagonalisation (or Lanczos at $T = 0$) [23] and the numerical renormalization group [180, 22].
For reviews on DMFT see e.g. [67, 81, 68, 105].

The general concept underlying a mean–field theory is the mapping of a lattice problem onto a single site problem by casting all non–local degrees of freedom, i.e., intersite couplings, into an effective *mean field* (compare Fig. 2.2 & 2.3). This mean field is then calculated with the request, that the single site problem reproduces the expectation values of the initial lattice problem. The identification of the single site observable with the local component of the corresponding lattice quantity is the actual approximation of a mean field approach. This procedure becomes exact in the limit of infinite coordination number (i.e. infinite dimensions) as a result of the central limit theorem (see below). For example, in the case of the "classical" Ising model

2.2 Dynamical Mean Field Theory

(Fig. 2.2) the coupling constant J has to be scaled with the inverse spatial dimension $1/d$ in order that the energy of the system remains finite and, hence, the deviation of the mean field from the true local field scales as $1/\sqrt{d}$, i.e., vanishes in the limit $d \to \infty$.

But let us rather turn to the scaling of Hamiltonians which we then want to combine with the ab initio methods in the limit of $d \to \infty$. Better yet, for more intuitive arguments this limit should be understood as the limit of infinite neighboring sites $\mathcal{Z}_{\|i-j\|} \to \infty$ (where $\|i-j\|$ is the distance between the sites). Turning to the Hubbard Hamiltonian (2.2) we can consider the proper scaling of the kinetic, and of the interaction part separately. In fact, the scaling of the latter one in the case of the Hubbard Hamiltonian is trivial: Since the interaction parameter U_{lmno} is defined as purely local this part of Hamiltonian (2.2) stays at a finite constant:

$$\left\langle \sum_{lmno\sigma\sigma'} U_{lmno} \hat{c}^\dagger_{il\sigma} \hat{c}^\dagger_{im\sigma'} \hat{c}_{in\sigma'} \hat{c}_{io\sigma} \right\rangle \xrightarrow{\mathcal{Z}_{\|i-j\|} \to \infty} \text{const.} \qquad (2.9)$$

where $\langle \mathcal{O} \rangle$ denotes the thermal average of $\mathcal{O}$.

For the first term of (2.2) the situation is obviously more involved and we have to take care that the kinetic energy *per site*, namely $\sum_{ljm\sigma} t_{il,jm} \hat{c}^\dagger_{il\sigma} \hat{c}_{jm\sigma}$ scales properly, i.e remains finite in $d \to \infty$. In this sum we have $\mathcal{Z}_{\|i-j\|}$ equivalent j terms. Hence, the kinetic energy per site i diverges unless we rescale $t_{il,jm}$ in the limit $\mathcal{Z}_{\|i-j\|} \to \infty$. To avoid this divergence, the following renormalization is considered:

$$t_{il,jm} = \frac{\tilde{t}_{il,jm}}{\sqrt{\mathcal{Z}_{\|i-j\|}}} \qquad (2.10)$$

with $\tilde{t}_{il,jm}$ staying constant upon increasing $\mathcal{Z}_{\|i-j\|}$. It can be easily shown, that the non–interacting Green function $G^0_{il,jm}(\omega)$ also scales just like $\tilde{t}_{il,jm}$ since they are directly connected:

$$G^0_{il,jm}(\omega) = \left[(\omega \underline{\underline{1}} - \underline{\underline{t}})^{-1}\right]_{il,jm} \propto \frac{1}{\sqrt{\mathcal{Z}_{\|i-j\|}}} \qquad (2.11)$$

2 Combining DFT and many-body approaches

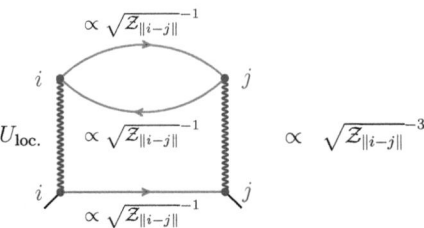

Fig. 2.4: Example for a second order diagram for the self energy (amputated legs). If we consider the scaling of the three Green functions connecting i and j shows that the contribution of this diagram vanishes in the DMFT limit of $d \to \infty$ and only the local contribution with $i = j$ survives.

where the double underlines denote matrices in the basis of orbital and site indices (we will keep this notation throughout the manuscript). Now, in order to see how the kinetic energy term scales, we need to know how the full Green function scales, since the scaling of $\langle \hat{c}_{il\sigma}^\dagger \hat{c}_{jm\sigma} \rangle$ is obviously directly connected to the scaling of the propagator $\langle \hat{c}_{jm\sigma} \hat{c}_{il\sigma}^\dagger \rangle$, i.e., the Green function $G_{il,jm}(\omega)$. Actually, we can show that it scales just like $G^0_{il,jm}(\omega)$. The way to do so, is to employ the Dyson equation and then show that the self energy is purely local in the limit of infinite coordination number.

In Feynman diagram notation we can write the Dyson equation for the Green functions as

$$\qquad = \qquad + \qquad \Sigma \qquad \qquad (2.12)$$

where $\Rightarrow$ denotes the full Green function, $\rightarrow$ the non–interacting one, and $\Sigma = \Sigma(\mathbf{k}, \omega)$ the full self energy, which is generally a function of momentum and frequencies. Rewriting the equation in k–space gives for the full Green function:

$$\underline{\underline{G}}(\mathbf{k}, \omega) = \left(\underline{\underline{G}}^0(\mathbf{k}, \omega)^{-1} - \underline{\underline{\Sigma}}(\omega, \mathbf{k}) \right)^{-1} \qquad (2.13)$$

2.2 Dynamical Mean Field Theory

Now consider the diagrams, that define the self energy. For instance, let us consider the case of the diagram shown in Fig. 2.4. Each $\rightarrowtail$ in this diagram connecting i and j with $i \neq j$ scales like $\sqrt{\mathcal{Z}_{\|i-j\|}}^{-1}$, which means that the total contribution of this diagram is $\sqrt{\mathcal{Z}_{\|i-j\|}}^{-3}$. But there is only a factor $\mathcal{Z}_{\|i-j\|}$ for the sum over the different j thus, the contribution scales like $\sqrt{\mathcal{Z}_{\|i-j\|}}^{-1}$ and becomes irrelevant in the limit $\mathcal{Z}_{\|i-j\|} \to \infty$. Moreover, this holds for any other diagram in which two sites $i \neq j$ are connected by three or more independent $\rightarrowtail$.

Yet, there are also diagrams, in which sites $i \neq j$ are connected by only two $\rightarrowtail$ (see Fig. 2.5 left hand side). These diagrams will contribute, since they scale like $\mathcal{Z}_{\|i-j\|}^{-1}$: Taking into account the factor $\mathcal{Z}_{\|i-j\|}$ for the sum over j the total contribution of these diagrams scales like 1. However, these diagrams are already taken into account if we substitute the non–interacting Green function by the full Green function like in Fig. 2.5 right hand side. In summary this means, that all diagrams contributing to the self energy are purely local, and that we can conclude from Eq. (2.12) eventually that:

$$G_{il,jm}(\omega) \propto \frac{1}{\sqrt{\mathcal{Z}_{\|i-j\|}}} \qquad (2.14)$$

so that:

$$\left\langle \sum_{iljm\sigma} t_{il,jm} \hat{c}^\dagger_{il\sigma} \hat{c}_{jm\sigma} \right\rangle \xrightarrow{\mathcal{Z}_{\|i-j\|} \to \infty} \text{const.} \qquad (2.15)$$

Note, the two factors $\sqrt{\mathcal{Z}_{\|i-j\|}}$ from the hopping and the Green function will just cancel the factor $\mathcal{Z}_{\|i-j\|}$ from the sum over j. This shows, that our initial ansatz (2.10) was correct. On the way, we also obtained the important result that the self energy becomes *purely local* in $d \to \infty$:

$$\Sigma_{ij}(\omega) \xrightarrow{\mathcal{Z}_{\|i-j\|} \to \infty} \delta_{ij} \Sigma(\omega) \qquad (2.16)$$

2 Combining DFT and many-body approaches

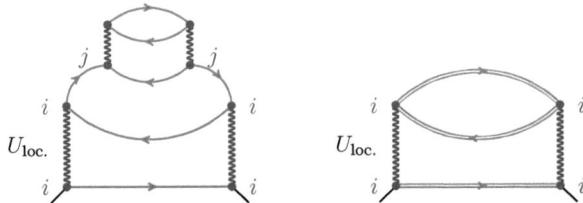

Fig. 2.5: Left hand side: Example of second order diagrams for the self energy (amputated legs) that will contribute in the $d \to \infty$ limit. Here i and j are connected only by two Green functions (straight lines). These kind of diagrams are contained in the *full local* diagram which is shown on the right hand side. a

or after a Fourier transform of the spatial variables:

$$\Sigma(\mathbf{k}, \omega) \xrightarrow{z_{\|i-j\|} \to \infty} \Sigma(\omega) \qquad (2.17)$$

Looking back, we have found a mean field theory in which the lattice problem is self consistently mapped onto a local problem (compare Fig. 2.3) just like in the mean field theory of Weiss for the Ising spins. And just like the Weiss mean field theory it becomes exact in the well defined limit of infinite coordination. Nonetheless, there is a significant difference to the classical mean field theory mentioned earlier: $\Sigma(\omega)$ and hence the local observable which depends on $\Sigma(\omega)$ are *frequency dependent*. Speaking in a pictorial way this means that we allow for the electrons on the local site to leave the site and travel around returning after some time, often denoted as "quantum fluctuations". This time dependence, in turn, yields the frequency dependence of our mean field Σ which makes it *dynamic*.

The final step missing, is the actual solution of the problem which means the determination of the local self energy. And just like in many other cases, this step is performed by mapping the new problem onto something which is well known and already solvable: the Anderson impurity model (AIM). There are several solvers available for the AIM ranging from analytical approximations like iterated perturbation theory (IPT) to numerically exact solvers like

Lanzcos diagonalisation or Quantum Monte Carlo (QMC) integration. The (self consistent) mapping of the DMFT problem to the AIM can be done since, diagrammatically, the DMFT self energy corresponds to the contribution of all topologically distinct local Feynman diagrams. Exactly the same diagrams can be obtained via the Anderson impurity model if its on-site interaction has the same form as the original Hamiltonian:

$$\hat{H}_{\text{AIM}} = \sum_{kl\sigma} \varepsilon_l(\mathbf{k}) \hat{a}^\dagger_{\mathbf{k},l,\sigma} \hat{a}_{\mathbf{k},l,\sigma} + \sum_{kl\sigma} \left[V_{lm}(\mathbf{k}) \hat{a}^\dagger_{\mathbf{k},l,\sigma} \hat{c}_{m,\sigma} + \text{h.c.} \right] + \sum_{ilmno\sigma\sigma'} U_{lmno} \hat{c}^\dagger_{il\sigma} \hat{c}^\dagger_{im\sigma'} \hat{c}_{in\sigma'} \hat{c}_{io\sigma} \quad (2.18)$$

Here the $\hat{a}^\dagger_{\mathbf{k},l,\sigma}$, $\hat{a}_{\mathbf{k},l,\sigma}$ are creators and annihilators of the non–interacting conduction band with dispersion relation $\varepsilon_l(\mathbf{k})$, the $\hat{c}^\dagger_{il\sigma}$, $\hat{c}_{il\sigma}$ are the creators and annihilators of the interacting impurity–site, and $V_{lm}(\mathbf{k})$ presents the hybridization between itinerant and localized electrons. If we now reformulate the problem in terms of a field–integral over Grassmann variables ψ and $\psi^\dagger$, we will see, that the conduction electrons can be integrated out. The propagator of the AIM reads:

$$G^\sigma_{lm}(i\omega_\nu) = -\frac{1}{\mathcal{Z}} \int \mathcal{D}\psi \mathcal{D}\psi^\dagger \left(\psi^\sigma_{\nu l} \psi^{\sigma\dagger}_{\nu m} e^{\mathcal{A}[\psi,\psi^\dagger,(\mathcal{G}^0)^{-1}]} \right) \quad (2.19)$$

where $\omega_\nu = \pi(2\nu+1)/\beta$ are the Matsubara frequencies defined with the inverse temperature $\beta = 1/k_\text{B}T$, the partition function:

$$\mathcal{Z} = \int \mathcal{D}\psi \mathcal{D}\psi^\dagger \left(e^{\mathcal{A}[\psi,\psi^\dagger,(\mathcal{G}^0)^{-1}]} \right) \quad (2.20)$$

and the single–site action:

$$\mathcal{A}\left[\psi,\psi^{\dagger},(\mathcal{G}^{0})^{-1}\right] = \sum_{\nu\sigma lm} \psi_{\nu m}^{\sigma\dagger} \left[\mathcal{G}_{lm}^{\sigma 0}(\mathrm{i}\omega_{\nu})\right]^{-1} \psi_{\nu l}^{\sigma} -$$
$$\sum_{lmno\sigma\sigma'} U_{lmno} \int_{0}^{\beta} d\tau \psi_{l}^{\sigma\dagger}(\tau)\psi_{n}^{\sigma'}(\tau)\psi_{m}^{\sigma'\dagger}(\tau)\psi_{o}^{\sigma}(\tau) \quad (2.21)$$

where the integral is performed over the imaginary time τ. The non–interacting Green function (Weiss field) of the AIM is given by:

$$\left[\mathcal{G}_{lm}^{\sigma 0}(\mathrm{i}\omega_{\nu})\right]^{-1} = \mathrm{i}\omega_{\nu} + t_{ikim} + \mu - \sum_{\mathbf{k}n} \frac{V_{nl}^{\dagger}(\mathbf{k})V_{nm}(\mathbf{k})}{\mathrm{i}\omega_{\nu} + \mu - \varepsilon_{n}(\mathbf{k})} \quad (2.22)$$

As already mentioned, the topology of the irreducible diagrams of this effective Anderson impurity model is exactly the same as the DMFT single site problem: simply the local contribution of all Feynman diagrams. Now we only have to identify the interacting Green function of the AIM with the local DMFT Green function in order to get identical self energies. This means, that we have to find the appropriate AIM, for which this equality holds. From the request $\underline{G}^{\mathrm{AIM}}(\omega) \equiv \underline{G}_{\mathrm{loc.}}^{\mathrm{DMFT}}(\omega)$ it follows directly, that the non–interacting AIM Green function:

$$\left[\underline{\mathcal{G}}^{0}(\mathrm{i}\omega_{\nu})\right]^{-1} = \left[\underline{G}(\omega)\right]^{-1} + \underline{\Sigma}(\omega) \quad (2.23)$$

With Eq. (2.23), which is basically a Dyson equation (compare to Eq. (2.13)) for the AIM, we have everything we need to build up a self–consistency scheme for the calculation of the local self energy:

1. Choose a starting self energy and calculate the local Green function by a k–integration of Eq. (2.13):

$$\underline{G}^{\mathrm{loc.}}(\omega) = \frac{1}{V_{\mathrm{BZ}}} \int_{\mathrm{BZ}} d^{3}k \frac{1}{(\omega + \mu)\underline{1} - \underline{\varepsilon}(\mathbf{k}) - \underline{\Sigma}(\omega)} \quad (2.24)$$

2. Then calculate the AIM Weiss field $\underline{\underline{\mathcal{G}}}^0$ via Eq. (2.23)

3. Solve the Anderson impurity model to obtain a new local Green function and self energy

4. compare new and old self energy and close the self consistency loop

This procedure is also sketched in Fig. 2.6. The DMFT method turned out to be a huge step forward for the understanding of many body systems by means of e.g. the Hubbard model or the periodic Anderson model and in particular improved the insight on the Mott–Hubbard metal-to–insulator transition [67]. Moreover, it also yielded new insights to the "dynamical features" of the Mott insulating phase by means of (spin)–polaronic side bands [181]. We shall now see, how the DMFT and its self consistent solution can be implemented in combination with the LDA in order to perform ab initio calculations of strongly correlated systems.

2.3 LDA+DMFT

In section 2.1 and 2.2 both the DFT(LDA) and the DMFT approaches were discussed with the motivation of merging the two methods into a single scheme, with which the *ab initio* treatment of strongly correlated electron systems becomes feasible.

The connection between LDA (or similar DFT schemes) and DMFT is most intuitive and straightforward on the Hamiltonian level. Therefore, the first step is to transform the LDA Hamiltonian (2.8) into a basis where we want to define the interaction part of the full Hamiltonian. We have to integrate out

2 Combining DFT and many-body approaches

INFO: AIM solver

The most difficult step in the DMFT self consistency is the solution of the impurity model in each iteration step for a new dynamic Weiss field $\underline{\mathcal{G}}^0(i\omega_n)$. Fortunately, there exist many well–established solvers for the quantum impurity model which have been developed over the last forty years. These solvers can be divided into analytical and numerical ones.

Analytical solvers are the iterated perturbation scheme (IPT) [67] and the noncrossing approximation method [95, 96, 161, 162]. However, when it comes to quantitative material calculations, in which we are interested here, the numerically exact solvers are preferable:

A straightforward way to solve the impurity–problem is the exact diagonalisation method [23, 132]. Here the AIM is solved with an approximated parameterized effective bath consisting of a few orbitals. The drawback of this technique is, however, that the number of orbitals one can effectively treat is severely limited by the size of an exponentially growing Hilbert space.

Another "numerically" exact method is the quantum Monte Carlo (QMC) method. One of the most common QMC impurity solvers for the DMFT is based on the Hirsch–Fye [85] algorithm (HF-QMC), which considers the single–impurity problem in discretized imaginary time. Within the HF–QMC the AIM is mapped onto a sum of noninteracting problems where a single particle moves in a fluctuating time dependent field. The effective bath only enters through the Weiss–field $\underline{\mathcal{G}}^0$, and there is no need to discretize the conduction band. With the QMC a much larger number of orbitals can be treated compared to the exact diagonalisation. Yet, there are also limitations to the QMC approach, although of a different nature: the computational effort grows as T^{-3}, so that this method is limited to high temperatures. In this sense the Lanczos diagonalisation and the QMC are somewhat complementary techniques. (there are however, schemes like the so called 'projective QMC' [216, 58] which overcome the high temperature restriction accessing the T= 0 regime).

By now there are other approaches, namely the so called continuous time QMC (CT-QMC) in its weak–coupling [175] and strong–coupling [215] versions which overcome the error of the Trotter discretization of the HF-algorithm. However, Blümer has very recently shown [18] that a state-of-the-art implementation of the HF-QMC is still competitive with the CT-QMC.

2.3 LDA+DMFT

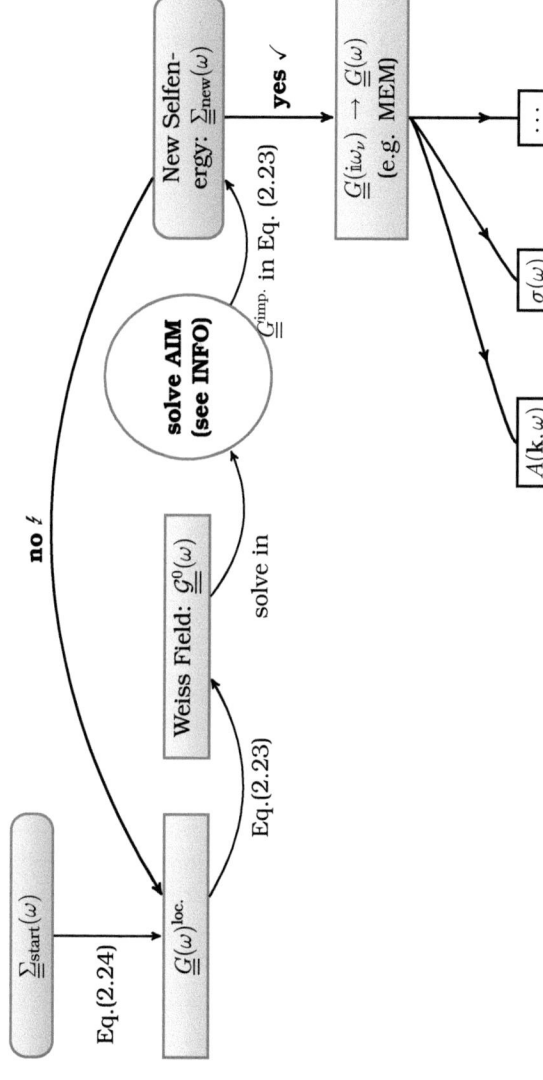

Fig. 2.6: DMFT self consistency cycle. Even though no rigorous proof exists concerning the convergence of the iterative process, practice has shown that it is usually not difficult to reach a self consistent solution of the scheme. Convergence is usually attained after a few iterations, depending on the number of degrees of freedom (i.e. orbitals) involved and also, more importantly depending on the closeness to transition points where one encounters a critical slowing down of the convergence.

2 Combining DFT and many-body approaches

subspaces of the full LDA Hamiltonian that are irrelevant for the excitations which we want to analyze, since the full LDA basis (i.e. of the order of $\mathcal{O}(100)$ orbitals) would be impossible to handle by means of any impurity solver. As already mentioned briefly at the end of section 2.1, maximally localized Wannier functions $\phi(\underline{r})$ [134] or wave functions of NMTO downfolding calculations [3] present an adequate choice for such a basis. In this basis we rewrite the wave functions:

$$\hat{\Psi}^{\sigma\dagger}(\underline{r}) = \sum_{il} \hat{c}^\dagger_{il\sigma} \phi^*_{il}(\underline{r}) \qquad (2.25)$$

and the Hamiltonian (2.8) reads:

$$\hat{H}^*_{\text{LDA}} = \sum_{ljm\sigma} t_{il,jm} \hat{c}^\dagger_{il\sigma} \hat{c}_{jm\sigma} \qquad (2.26)$$

Here the asterix denotes the projection/downfolding character of the basis $\phi_{il}(\underline{r})$ and the "tight binding" hopping terms read

$$t_{il,jm} = \int d^3r \phi^*_{il}(\underline{r}) \left[-\frac{\hbar^2}{2m_e}\Delta + V_{\text{ion}}(\underline{r}) + \int d^3r' V_{\text{ee}}(\underline{r}-\underline{r}')\rho(\underline{r}') + \frac{\partial E^{\text{LDA}}_{\text{xc}}(\rho)}{\partial \rho(\underline{r})} \right] \phi^*_{jm}(\underline{r}) \qquad (2.27)$$

Now, if we compare the Hubbard Hamiltonian (2.2) with the LDA Hamiltonian in the basis of localized Wannier functions (2.26) it is straightforward to identify the kinetic part of the Hubbard Hamiltonian with the LDA Hamiltonian on the Wannier basis (for which we know the solution $t_{il,jm}$). Hence, after finding an appropriate expression for the $\underline{U}$ matrix of the interacting part of the Hubbard Hamiltonian, the problem (2.28) can be directly solved by the DMFT self consistency (Fig. 2.6).

$$\hat{H}_{\text{eff.}} = \hat{H}^*_{\text{LDA}} + \hat{H}_{\text{int.}}(\underline{U}) \qquad (2.28)$$

2.3 LDA+DMFT

Specifically, the projected/downfolded LDA Hamiltonian enters in its Fourier–transformed form as $\underline{\varepsilon}(\mathbf{k}) = \underline{\varepsilon}^{\text{LDA}}(\mathbf{k})$ in Eq. (2.24)

An important remark should be made at this point: It must not be forgotten, that the LDA scheme already treated a part of the electronic Coulomb interaction, e.g. the static Hartree part of it: $\int d^3r' V_{\text{ee}}(\underline{r} - \underline{r}')\rho(\underline{r}')$- *These terms must not be double counted!* To avoid this, we have to introduce the so called "double counting correction" (DC) in our scheme. Throughout this work the DC was performed along the lines of Anisimov [9, 81].

$$\Delta\varepsilon = \bar{U}\left(n_L - \frac{1}{2}\right) \quad (2.29)$$

where n_L denotes the LDA density of the orbital subspace L (e.g. either 3d, 2p, etc.) and $\bar{U}$ is the average of the interaction matrix which, in a cubic symmetry (including also a Hund's coupling J), can be written for M interacting orbitals as

$$\bar{U} = \frac{U + (M-1)(U-2J) + (M-1)(U-3J)}{2M-1} \quad (2.30)$$

Still, the question of the most appropriate choice for the DC term is an ongoing debate. It should be noted, that the Anisimov style DC only becomes relevant, when $\hat{H}^*_{\text{LDA}}$ includes subspaces of different angular–momentum quantum number from the full LDA Hamiltonian. If we only take into account "effective" d–states only, the DC just corresponds to a shift of the total energy which can be absorbed by the chemical potential. We refer to chapter 5 where we formulate models on extended basis sets on which the corresponding DC has to be taken into account explicitly.

Before closing this methodological chapter with an overview of the observables that can be calculated within LDA+DMFT for comparison with experiment, a brief comment concerning the choice of the appropriate subspace to formulate $\hat{H}^*_{\text{LDA}}$ and $\hat{H}_{\text{int.}}(\underline{U})$ is due.

In summary, we have seen how *ab initio* methods could be used in order to construct an interacting Hamiltonian (2.28) which, in turn, is solved by the

2 Combining DFT and many-body approaches

procedure sketched in Fig. 2.6. In a way, we introduce the many–body correlation effects "on top" of the LDA results. With this approach comes a certain risk: What if degrees of freedom, that were projected out in the first place would have become important *if* they would have been included. In other words: What is the true minimal basis–set in order to capture all features correctly?[3]

Today, much effort is put into finding a closed LDA+DMFT self consistency on a full basis–set. The first step towards such implementations is to expand the basis–set of the self–consistency but treat only certain subspaces (i.e. the ones where correlations become essential) with DMFT and others with less sophisticated methods. For further discussions concerning this important topic, we refer to chapter 5.

2.3.1 Obtaining the local self energy on the real axis

Employed as DMFT impurity solver, the most quantum Monte Carlo simulations produce Green functions $G(\tau)$ of imaginary time $\tau = it$. However, real–frequency results are crucial since most experiments probe dynamical quantities like spectral functions, etc. Thus, the inability to extract real-frequency or real–time results from Matsubara (imaginary) time QMC simulations poses the necessity of an additional postprocessor to perform the analytical continuation.

The relation between $G(\tau)$ and $A(\omega) = -\frac{1}{\pi}\Im(G(\omega))$ is, in fact, linear and surprisingly simple:

$$G(\tau) = \int d\omega \mathcal{K}(\tau,\omega) A(\omega) \qquad (2.31)$$

Nevertheless, inversion is complicated by the exponential nature of the kernel. For example, for a fermionic single particle Green function the kernel

---

[3]For example it is easy to realize that the physics of a d/p charge–transfer insulator cannot be captured entirely by a model that only includes d degrees of freedom - see sections 3.2 and 5.1.2

reads [182]

$$\mathcal{K}(\tau,\omega) = \frac{e^{-\tau\omega}}{1+e^{-\beta\omega}} \qquad (2.32)$$

For finite τ and large ω the kernel is exponentially small, so that $G(\tau)$ is insensitive to the high frequency features of $A(\omega)$. Equivalently, if we approximate both G and A by equal–length vectors and $\mathcal{K}$ by a square matrix, then we find that the determinant of $\mathcal{K}$ is exponentially small, so that $\mathcal{K}^{-1}$ is ill–defined. This means, that there are an infinite number of A with very different characters (it might be even not causal) that yield the same G.

In cases where G is extremely precise the so called Padé method [209] can be employed, where G is fit to a functional form, usually the ratio of two polynomials. This can then be easily analytically continued by replacing $i\omega_n \to \omega + i0^+$. However, QMC data of G is usually far from being suitable for this kind of technique. Here the most appropriate approaches turned out to be entropy–based methods of Bayesian data analysis known as *Maximum Entropy Method* or short MEM. In the MEM we do not ask "What is the spectrum to the G we measured?", but rather "Which A maximizes the *a posteriori* probability given the data of G?". In order to do so, $A(\omega)$ is interpreted as a Probability density on an interval $[-\omega_0/2, \omega_0/2]$. The corresponding entropy then reads:

$$S = \int_{-\omega_0/2}^{\omega_0/2} d\omega A(\omega) \ln(A(\omega)\omega_0) \qquad (2.33)$$

The prior probability for A is then [187] proportional to $P(A) \propto \exp(\alpha S)$ where α is a parameter which should maximize $P(A)$. Moreover, $A(\omega)$ has to yield the correct $G(\tau)$. The likelihood function for this is

$$P(G|A) = \exp(-\frac{1}{2}\chi^2) \qquad (2.34)$$

where χ^2 is the square deviation of the $G(\tau)$ calculated from $A(\omega)$ compared to the measured $G(\tau)$. Now the *a posteriori* probability of $A(\omega)$ yielding $G(\tau)$ is

2 Combining DFT and many-body approaches

$$P(A|G) = P(G|A)P(A)/P(G) \propto \exp(\alpha S - \frac{1}{2}\chi^2) \qquad (2.35)$$

which is maximized in the procedure. (During the procedure we work with one set of QMC data at a time so that $P(G)$ is constant and may be dropped.) For further reading and (many) details we refer to the exhaustive review by Jarrell and Goubernatis [94].

2.4 Observables from LDA+DMFT

Up to this point the merger of *ab initio* DFT methods (and specifically the LDA) with the many–body DMFT approximation have been introduced. The motivation, from the DFT perspective, was the inclusion of electronic correlations which are out of reach for simple LDA (or even LDA+U) calculations. On the other hand, the motivation from the many–body perspective was to leave the territory of idealized model Hamiltonians and construct realistic, yet solvable, Hamiltonians. These motivations are aimed at a direct comparison between theory and experiment. Hence, a discussion of the observables which allow for such comparison is due.

A thorough understanding of what is actually measured in an experiment and how it is related to the calculated quantities is extremely important. Usually the processes in an experimental measurement are quite complicated and when the data should be compared to the calculations, certain approximations have to be made. However, it is essential to know what approximations are justified at each step of the way.

In the following sections three of the most common spectroscopic techniques and their comparison to LDA+DMFT calculations will be discussed shortly: the angular resolved photoemission spectroscopy ARPES, optical conductivity IR measurements, and finally the X–ray absorption spectroscopy XAS. We

2.4 Observables from LDA+DMFT

shall meet all of these techniques again in the following chapters when we refer to experimental results.

2.4.1 Angular Resolved Photoemission Spectroscopy (ARPES)

One of the most popular spectroscopic tools is the angular resolved photoemission (ARPES). In a PES experiment, the probe light induces an emission of electrons from the sample. The emitted electrons are then analyzed by means of their energy.

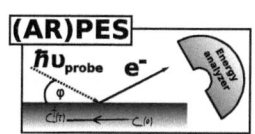

We will see that, in a many body context, PES should be seen as the measurement of a *"quasihole" propagation*. One reason for the great popularity of this approach is the fact that its spectra can be related to single particle propagators, i.e., Green functions of the system, which are a direct result of many theoretical calculations. Also the LDA+DMFT self consistent scheme, as it was discussed in the previous section, yields the (local) Green function directly:

Let $|j\rangle$ and $|i\rangle$ be eigenstates of a general (specifically also interacting) Hamiltonian with $\hat{H}|j\rangle = E_j|j\rangle$. Then the Green function on the imaginary Matsubara axis $i\omega_\nu$ and the definition of the so called spectral function $\underline{A}(\mathbf{k},\omega)$ reads

$$
\begin{aligned}
G_{m,n}(\mathbf{k}, i\omega_\nu) &= -\frac{1}{\mathcal{Z}} \int_0^\beta d\tau e^{i\omega_\nu \tau} \operatorname{Tr}\left[e^{-\beta \hat{H}} e^{\hat{H}\tau} \hat{c}_{n\mathbf{k}} e^{-\hat{H}\tau} \hat{c}_{m\mathbf{k}}^\dagger\right] \\
&= -\frac{1}{\mathcal{Z}} \int_0^\beta d\tau e^{i\omega_\nu \tau} \sum_{i,j} e^{-\beta E_i} e^{(E_i - E_j)\tau} \langle i|\hat{c}_{m\mathbf{k}}|j\rangle \langle j|\hat{c}_{n\mathbf{k}}^\dagger|i\rangle \\
&= -\frac{1}{\mathcal{Z}} \sum_{i,j} e^{-\beta E_i} \frac{e^{(i\omega_\nu + E_i - E_j)\beta} - 1}{i\omega_\nu + E_i - E_j} \langle i|\hat{c}_{m\mathbf{k}}|j\rangle \langle j|\hat{c}_{n\mathbf{k}}^\dagger|i\rangle \\
&\equiv \int d\omega' \frac{A_{m,n}(\mathbf{k}, \omega')}{i\omega_\nu - \omega'}
\end{aligned}
\qquad (2.36)
$$

where m and n denote the spin and orbital quantum numbers, $\mathcal{Z}$ is the grand canonical partition function, and $\beta = 1/k_\mathrm{B}T$. So that

2 Combining DFT and many-body approaches

$$A_{m,n}(\mathbf{k},\omega) = \frac{1}{\mathcal{Z}}\left(1 \pm e^{-\beta\omega}\right)\sum_{i,j} e^{-\beta E_i} \langle i|\hat{c}_{m\mathbf{k}}|j\rangle\langle j|\hat{c}^{\dagger}_{n\mathbf{k}}|i\rangle\delta(\omega + E_i - E_j) \quad (2.37)$$

For the orbital diagonal part ($m = n$), we can relate $A_{n,n}(\mathbf{k},\omega)$ directly to the retarded single particle Green function:

$$A_{n,n}(\mathbf{k},\omega) = -\frac{1}{\pi}\Im G^{\text{ret.}}_{n,n}(\mathbf{k},\omega) \quad (2.38)$$

Further, it can be shown (see e.g. [214]), that the ARPES intensity in the *sudden approximation* can be related to $A(\mathbf{k},\omega)$ by

$$I^{\text{ARPES}}(\mathbf{k},\omega) = \frac{2\pi}{\hbar}\Delta(\mathbf{k})f(T,\omega)A(\mathbf{k},\omega) \quad (2.39)$$

where $f(T,\omega)$ is the Fermi distribution and $\Delta(\mathbf{k})$ the so called transition matrix element which, generally a function of **k**, is very often approximated to be constant. To sum it up in a nutshell, ARPES probes the propagation of an electron–hole created by absorption of a photon. Hence, in the Fermi liquid regime the ARPES spectra around the Fermi energy can be understood as the renormalized free–particle spectrum[4], i.e., the propagation of a "quasi hole".

---

[4]It is common slang to say that ARPES measures the "bandstructure" or the "occupied DOS" – strictly speaking this statement is incorrect. A spectrum is never a DOS, although it can be interpreted in this way in a Fermi liquid context. The fact that the integral over the spectrum is related to the number of electrons just reflects a *sum–rule*

2.4.2 Optical Conductivity (IR)

The next technique to be discussed is the response of the system to electromagnetic radiation in the visible/infrared part of the spectrum – the so called optical conductivity. The experiment is often set up in a geometry that allows for the measurement of the reflectivity.

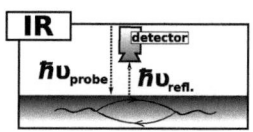

From the reflectivity one computes, via Kramers Kronig (KK) transformation, the optical conductivity which is more transparent in its physical interpretation[5]. While in ARPES the photon energy was large enough to actually "kick out" an electron, the excitations of IR measurements are true two–particle (electron/hole) excitations at a constant particle number.

Technically this means, that we have to calculate two–particle Green functions in order to capture the IR spectra. In fact, we will see later that in the specific case of the DMFT approximation the IR spectra can be obtained from the one–particle spectral functions. Nonetheless, this is a good point to introduce the concept of calculating of response functions within quantum field approaches such as DMFT generally.

The conductivity tensor $\underline{\sigma}$ is defined as the response function relating the "perturbation" of an electric field $\underline{E}$ with the systems response, the current $\underline{J}$, in Ohm's law

$$J_\alpha(\underline{r},t) = \int d^3r \int_{-\infty}^{t} dt' \sigma_{\alpha\beta}(\underline{r},\underline{r}',t-t') E_\beta(\underline{r}',t') \qquad (2.40)$$

with α and β being the Cartesian coordinates (x,y,z) and $\underline{J}$ the expectation value of the current density operator $\hat{j}$

$$\underline{J}(\underline{r},t) = \langle \hat{j}(\underline{r},t) \rangle \qquad (2.41)$$

which we will derive by means of the linear response theory in the following, and the electric field

---

[5]In order to perform the KK transformation the experimental data has to be extrapolated to zero and infinite frequency – for details see for example [217, 51]

2 Combining DFT and many-body approaches

$$\underline{E}(\underline{r},t) = \underline{E}_0 \, e^{i(\mathbf{q}\underline{r}-\omega t)} \tag{2.42}$$

which we assume, for simplicity, to consist of just one single mode $(\mathbf{q}, \omega)$. Further we can argue that for IR experiments the wavelength of the radiation in the optical spectrum is quite large compared to the typical length scales (i.e. atomic distances and penetration depth) of the solid. Hence we neglect spatial variations of the field and, accordingly, conclude that the response at position $\underline{r}$ to the field $\underline{r}'$ only depends on the distance $\underline{r} - \underline{r}'$. Then we can write, after Fourier transform to momentum and frequency space

$$\underline{J}(\mathbf{q},\omega) = \underline{\sigma}(\mathbf{q},\omega)\underline{E}(\mathbf{q},\omega) \tag{2.43}$$

and take the limit of zero momentum transfer $\mathbf{q} \to 0$, which, in fact, corresponds to the dipole approximation. In these terms the optical conductivity refers to the real part of the conductivity tensor in the $\mathbf{q} = 0$ limit[6]: $\Re\underline{\sigma}(\mathbf{q}=0, \omega)$. The limit $\omega \to 0$ will result in the direct current (dc) conductivity. Let us now turn to the actual calculation of the optical conductivity. The first quantity to derive is the current density operator $\hat{\jmath}$ which is defined from the coupling with the electromagnetic vector potentials. It is convenient here to divide the electronic Hamiltonian into a non–interacting and an interacting part

$$\hat{H} = \hat{H}_0 + \hat{H}_{\text{int.}} \tag{2.44}$$

where $\hat{H}_{\text{int.}}$ consists of the two–particle operators of the Coulomb interaction or the coupling to bosonic modes, and $\hat{H}_0$ can be formulated with the generalized momentum (minimal substitution) in an electromagnetic field as

$$\hat{H}_0 = \sum_\sigma \int d^3r \, \hat{\Psi}^{\sigma\dagger}(\underline{r},t) \left[\frac{1}{2m_e}\left(-i\hbar\Delta - \frac{e}{c}\underline{A}(\underline{r},t)\right)^2 + V(\underline{r},t) \right] \hat{\Psi}^\sigma(\underline{r},t) \tag{2.45}$$

---

[6] at this point it is easy to realize that the gap of an optical spectrum, which corresponds to the smallest $\Delta\omega$ of a direct transfer, can differ severely from e.g. a PES gap

2.4 Observables from LDA+DMFT

Here, $\underline{A}(\underline{r},t)$ is the vector potential and c is the speed of light. We work in the Coulomb gauge assuming charge neutrality of the system so that $V(\underline{r},t) = 0$. We derive the current density operator to be

$$\hat{\jmath}(\underline{r},t) = -\frac{i\hbar}{2m_e}\sum_\sigma \left(\hat{\Psi}^{\sigma\dagger}(\underline{r},t)\Delta\hat{\Psi}^\sigma(\underline{r},t) - (\Delta\hat{\Psi}^{\sigma\dagger}(\underline{r},t))\hat{\Psi}^\sigma(\underline{r},t)\right) + \frac{e^2}{m_e c}\underline{A}(\underline{r},t)\rho(\underline{r},t) \quad (2.46)$$

With (2.46) we can rewrite (2.45) as

$$\hat{H}_0 = -\frac{1}{c}\int d^3r\,\hat{\jmath}(\underline{r},t)\underline{A}(\underline{r},t) + \int d^3r\,\rho(\underline{r},t)V(\underline{r},t) \quad (2.47)$$

and identify, in turn, the current density as

$$\hat{\jmath}(\underline{r},t) = -c\frac{\delta\hat{H}_0}{\delta\underline{A}(\underline{r},t)} \quad (2.48)$$

Note that the minimal substitution $(-i\hbar\Delta \to (-i\hbar\Delta - \frac{e}{c}\underline{A}(\underline{r},t))$ has no effect on the interacting part of Hamiltonian (2.44) as long as it consists of density-density type of interactions. Now, turning back to the expression (2.46) for the current density, we can identify the two summands

$$\hat{\jmath}(\underline{r},t) = \hat{\jmath}_P(\underline{r},t) + \hat{\jmath}_D(\underline{r},t) \quad (2.49)$$

as the paramagnetic

$$\hat{\jmath}_P(\underline{r},t) = -\frac{i\hbar}{2m_e}\sum_\sigma \left(\hat{\Psi}^{\sigma\dagger}(\underline{r},t)\Delta\hat{\Psi}^\sigma(\underline{r},t) - (\Delta\hat{\Psi}^{\sigma\dagger}(\underline{r},t))\hat{\Psi}^\sigma(\underline{r},t)\right) \quad (2.50)$$

and diamagnetic term

$$\hat{\jmath}_D(\underline{r},t) = \frac{e^2}{m_e c}\underline{A}(\underline{r},t)\hat{\rho}(\underline{r},t) \quad (2.51)$$

2 Combining DFT and many-body approaches

With this separation, the expectation value of $\hat{j}$ becomes

$$\underline{J}(\underline{r},t) = \langle \hat{j}_P(\underline{r},t) \rangle + \frac{ne^2}{m_e c} \underline{A}(\underline{r},t) \qquad (2.52)$$

since, in linear response theory, $\langle \rho(\underline{r},t) \rangle = n$ for the system without the perturbation $\underline{A}$.

Looking at at equation (2.52), we define the kernel $\underline{K}(\underline{r}-\underline{r}',t-t')$

$$\underline{K}(\underline{r}-\underline{r}',t-t') \equiv \underbrace{\frac{e^2}{c}\underline{\chi}(\underline{r}-\underline{r}',t-t')}_{\text{paramag. kern.}} - \underbrace{\frac{e^2 n}{cm_e}\underline{1}}_{\text{diamag. kern.}} \qquad (2.53)$$

in order to write

$$\underline{J}(\underline{r},t) = \int d^3r \int_{-\infty}^{\infty} dt'\, \underline{K}(\underline{r}-\underline{r}',t-t')\underline{A}(\underline{r}',t') \qquad (2.54)$$

Yet, we started with the motivation to calculate the response to an electric field $\underline{E}$, like in equation (2.40), and not the vector potential $\underline{A}$. But, in fact we can relate those two quantities in the chosen gauge by

$$\underline{E} = -\frac{1}{c}\partial_t \underline{A} \qquad\qquad \underline{A}(\mathbf{q},r) = \frac{c}{i\omega}\underline{E}(\mathbf{q},\omega) \qquad (2.55)$$

Now, we can revisit equation (2.43) and, with the help of (2.55), (2.53), and (2.54) identify the conductivity tensor

$$\underline{\sigma}(\mathbf{q},\omega) = \frac{1}{i\omega}\underline{K}(\mathbf{q},\omega) \qquad (2.56)$$

In order to express the optical conductivity $\Re\underline{\sigma}(\mathbf{q},\omega)$, we rewrite (2.56) remembering that in the linear response theory the kernel, as a consequence of causality, is analytic in the upper half plane $\omega + i\delta$ and hence

2.4 Observables from LDA+DMFT

$$\underline{\sigma}(\mathbf{q},\omega) = \frac{1}{\mathrm{i}(\omega+\mathrm{i}\delta)}\underline{K}(\mathbf{q},\omega) = \mathcal{P}\left[\frac{1}{\mathrm{i}\omega}\underline{K}(\mathbf{q},\omega)\right] - \pi\delta(\omega)\underline{K}(\mathbf{q},\omega) \qquad (2.57)$$

So we can separate the real and imaginary part into

$$\Im\underline{\sigma}(\mathbf{q},\omega) = -\mathcal{P}\left[\frac{1}{\omega}\Re\underline{K}(\mathbf{q},\omega)\right] \qquad (2.58)$$

$$\Re\underline{\sigma}(\mathbf{q},\omega) = \mathcal{P}\left[\frac{1}{\omega}\Im\underline{K}(\mathbf{q},\omega)\right] - \pi\delta(\omega)\Re\underline{K}(\mathbf{q},\omega) \qquad (2.59)$$

and substitute (2.53)

$$\Re\underline{\sigma}(\mathbf{q},\omega) = \underbrace{\frac{e^2}{c\,\omega}\Im\underline{\chi}(\mathbf{q},\omega)}_{\text{regular part}} - \underbrace{\pi\delta(\omega)\left[\Re\underline{\chi}(\mathbf{q},\omega) - \frac{ne^2}{mc}\underline{1}\right]}_{\text{singular part}} \qquad (2.60)$$

Usually the singular part, whose presence would imply zero resistivity, is zero for normal materials. This is equivalent to the statement that the diamagnetic part of the current density will not contribute to the optical conductivity (exceptions are e.g. the superconductors where the singular part is not zero). Hence, let us neglect the second summand in (2.60) and calculate the regular part. Instead of giving a rigorous derivation of the final formula, we state it here giving some arguments and refer to [201, 1] for details. First of all the Kubo formula of the linear response theory will relate χ to the current–current correlation function which we write in k–space and on the imaginary time τ axis as

$$\chi_{\alpha\beta}(\mathbf{q},\tau) \propto \langle \mathcal{T} j_\alpha(-\mathbf{q},\tau) j_\beta(\mathbf{q},0) \rangle \qquad (2.61)$$

Here implicitly the homogeneity of the system and continuity of time is assumed. In the dipole approximation we are also only interested in the limit $q \to 0$ in which the current operator can be written on a Wannier–like localized basis set (denoted by the set of quantum numbers "L") as

2 Combining DFT and many-body approaches

$$j_\alpha(\mathbf{q}=0,\tau) = e \sum_{\mathbf{k},LL',\sigma} v^{LL'}_{\mathbf{k},\alpha} \hat{c}^\dagger_{\mathbf{k}L'\sigma}(\tau)\hat{c}_{\mathbf{k}L'\sigma}(\tau) \qquad (2.62)$$

The $v^{LL'}_{\mathbf{k},\alpha}$ are the elements of the Fermi velocity matrix at $\mathbf{q} = 0$. Thus,

$$\chi_{\alpha\beta}(\mathbf{q}=\underline{0},\tau) \propto \mathrm{i}\,\langle T j_\alpha(-\mathbf{q}=\underline{0},\tau) j_\beta(\mathbf{q}=\underline{0},0)\rangle$$

$$\propto \sum_{\mathbf{k}\bar{\mathbf{k}},LL',\bar{L}\bar{L}',\sigma\bar{\sigma}} v^{LL',\sigma}_{\mathbf{k},\alpha} v^{\bar{L}\bar{L}',\bar{\sigma}}_{\bar{\mathbf{k}},\beta} \left\langle T \hat{c}^\dagger_{\mathbf{k}L'\sigma}(\tau)\hat{c}_{\mathbf{k}L\sigma}(\tau)\hat{c}^\dagger_{\bar{\mathbf{k}}\bar{L}'\bar{\sigma}}(0)\hat{c}_{\bar{\mathbf{k}}\bar{L}\bar{\sigma}}(0)\right\rangle \qquad (2.63)$$

$$\propto \sum_{\mathbf{k},LL',\bar{L}\bar{L}',\sigma} v^{LL',\sigma}_{\mathbf{k},\alpha} G^{L\bar{L}',\sigma}_{\mathbf{k}}(\tau) v^{\bar{L}\bar{L}',\sigma}_{\mathbf{k},\beta} G^{\bar{L}L',\sigma}_{\mathbf{k}}(-\tau)$$

where in the last step we made several assumptions: First of all we contracted the four fermionic operators according to the Wick theorem and to one–particle Green functions. Next, we assumed the Green functions to be diagonal in the spin index (i.e. paramagnetic case). And finally we neglected so called "vertex corrections" which will now be discussed a little more detailed.

Diagrammatically the last expression in (2.63) corresponds to the bubble shown on the left hand side of Fig. 2.7. Turning to the neglected two particle vertex Γ we would have to sum up more terms including summands of the form shown in the middle of Fig. 2.7 which means excitonic particle–hole interactions. These terms could be written in frequency space (dropping orbital indices for simplicity)

$$\sum_{\mathbf{k},\omega_\nu,\sigma}\sum_{\mathbf{k}',\omega'_\nu,\sigma'} v^\sigma_{\mathbf{k}} G^\sigma(\mathbf{k},\mathrm{i}\omega_\nu) G^\sigma(\mathbf{k},\mathrm{i}\omega_\nu+\mathrm{i}\Omega_\nu)$$

$$\Gamma^{\sigma\sigma'}_{\mathbf{k},\mathbf{k}'}(\mathrm{i}\omega_\nu,\mathrm{i}\omega'_\nu,\mathrm{i}\Omega_\nu) G^{\sigma'}(\mathbf{k}',\mathrm{i}\omega'_\nu) G^{\sigma'}(\mathbf{k}',\mathrm{i}\omega'_\nu+\mathrm{i}\Omega_\nu) v^{\sigma'}_{\mathbf{k}'} \qquad (2.64)$$

The approximation we make by neglecting such terms can interestingly be understood in the DMFT limit of infinite coordination. It can be shown that in this limit the vertex can be replaced by a purely local quantity [223] so

2.4 Observables from LDA+DMFT

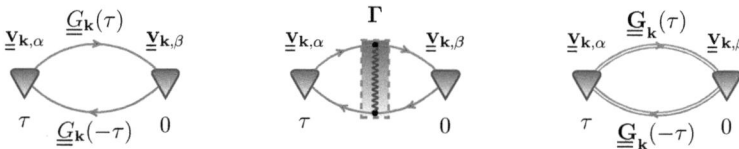

Fig. 2.7: Feynman diagrams for the optical conductivity (double underlines denote the matrix character in orbital and spin space). Left hand side: example of a bubble in the sum of Eq. (2.63). Center: Neglected excitonic diagrams with two particle vertex Γ. Right hand side: Bubble taking into account the *full* Green function

that the momentum sums in (2.64) can be performed individually for k and k' respectively. However, in the single-band case, the Fermi velocities are odd functions with respect to momentum, while the Green function is even so that eventually the sums become zero. Hence, in the DMFT limit the excitonic vertex corrections vanish for the single orbital case. In multi-orbital systems it is generally a question of symmetry if the same arguments hold – if not, it should be realized that the neglect of vertex corrections is approximative even in the limit of infinite coordination.

Finally, we exchange the non–interacting Green functions in the bubble by the fully interacting ones which is depicted in Fig. 2.7 on the right hand side. Eventually, after Fourier transformation to frequencies and analytical continuation to the real axis, we obtain for the optical conductivity

$$\Re\sigma^{\alpha\beta}(\Omega) = \frac{2\pi e^2 \hbar}{V} \sum_{\mathbf{k}} \int d\omega' \frac{f(\omega') - f(\omega' + \Omega)}{\Omega} \,\text{Tr}\left[\underline{v}_\alpha(\mathbf{k})\underline{\underline{A}}(\mathbf{k},\omega')\underline{v}_\beta(\mathbf{k})\underline{\underline{A}}(\mathbf{k},\omega'+\Omega)\right]$$
(2.65)

where f is the Fermi distribution, $\underline{\underline{A}}$ is the spectral function (not to be confused with the vetor potential) as defined in (2.39), and the double lines denote the matrix character in spin and orbital space. Equation (2.65) shows the remarkable fact, that the interaction part of the Hamiltonian (2.44) only enters in the evaluation of the spectral function $A(\mathbf{k},\omega)$. The Fermi velocity $\underline{v}_\alpha(\mathbf{k})$, on the other hand, is *independent* of the interactions.

2.4.3 X-ray Absorption Spectroscopy (XAS)

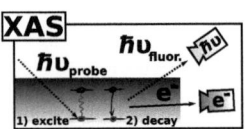

The last spectroscopic technique important for this work is the X-ray absorption spectroscopy. The process of XAS is the excitation of a core–shell electron with associated selection rules. Sometimes the XAS is said to measure the "unoccupied density of states" or in the case of correlated systems the "propagation of an additional electron" with the picture in mind that the core–electron plays the role of the added electron in the system. This picture would correspond to a kind of "inverse PES". However, in general this claim is not quite correct. For many XAS spectra the most prominent features can only be identified by excitations from the ground state to a very localized electron–core hole pair, a so called exciton. This final state has to be understood as a generic many–body excitation. The XAS excitation is, from a technical point of view, more related to the optical conductivity discussed in the previous section than to single particle spectroscopy like ARPES. In fact, it can be seen as complementary with respect to the IR spectroscopy . In our approximation (which becomes, for a single band case, even exact in the DMFT limit of infinite coordination) for the optical conductivity, we calculated the bubble on the right hand side of Fig. 2.7 which means the simultaneous propagation of an electron and a hole and neglected the excitonic electron–hole interactions between both. In XAS, on the other hand, these excitonic electron–(core)hole interactions are the most important energy scale since the core hole potential localizes the electron–hole pair so strongly that k is not a good quantum number anymore (i.e. there is no "propagation"). Hence, the most valuable information in XAS is best understandable in a local picture. For the transition metals the most informative excitations are the $2p$ to $3d$ transitions, the so called L–edges (at energies around ~ 350 eV to ~ 950 eV – i.e. soft x–ray range). These transitions are dipole allowed and therefore have large cross-sections. Yet, complementary valuable information can also be extracted from other core electron excitations like $1s$ to $3d$ (pre–edge region of the K–edge ~ 5 keV to ~ 9.5 keV – i.e. hard x–ray range). One of the remarkable advantages of XAS

2.4 Observables from LDA+DMFT

is the possibility of an element selective measurement, since the absorption edges have distinct energies due to the fact that the core shell electrons have element–specific bonding energies.

Further, more advanced techniques are available which make use of the polarization dependence of the absorption. Measuring the dependency of the absorption for circular polarized light, called circular dichroism (CD), is a spectroscopic way of measuring the susceptibility of a compound [76]. The dependency of the absorption for linear polarized light, called linear dichroism (LD), on the other hand has been proven to be a valuable tool for detection of orbital occupation [73, 76] as will be seen in chapter 3.

Important for the application of X–ray absorption as a spectroscopic technique is the development of models to simulate accurately the observed spectra: a quantitative analysis allows for the determination of various (near) ground state properties, including the valence, spin and orbital state of the atoms under investigation [60]. These models are based on cluster models (for an overview see e.g. [73, 76]) and have been successfully extended [198, 48, 195] as to include the influence of the core–hole using the full atomic multiplet theory. The scheme relies on the localized nature of the exciton and consists basically in diagonalizing a configuration interaction (CI) Hamiltonian. The calculations are parameter–based and, hence, not *ab initio*. Mostly, the parameters are fitted to experimental data in order to extract information. The "cluster calculations", as they are often called, became quite popular (especially for the L– and M–edges) due to their – sometimes outstanding – ability to reproduce experimental data. *Ab initio* methods like LDA often reproduce well the continuum part of the spectra or the main edge region of K–edges. However these methods fail if excitonic features become important (like in the L–edges of the pre–edge region of the K–edges), where for example non–single slater determinant states, i.e., states which no longer have the single particle nature, become important.

Nonetheless, recently DFT ground state results have been used successfully in order to calculate parameters needed in the CI calculations [78]. In this work we extend this philosophy to the application of a combination of LDA+DMFT with the CI calculations. Details will be discussed in the respec-

2 Combining DFT and many-body approaches

tive sections of chapter3. A nice review concerning which theoretical scheme is suited for which absorption edges can be found in [165]. For more general informations on XAS and CI calculations see [73, 76] and references therein.

3 Bulk 3d-transition metal compounds: Theory vs. Experiment

In the previous chapter the combination of *ab initio* DFT(LDA) calculations with the DMFT many body approach was discussed. Deriving low energy Hamiltonians from LDA bandstructure results and solving them by means of DMFT turned out to be an extremely successful method in order to perform realistic calculations for strongly correlated materials. Whereas the parameter model Hamiltonian calculations allowed only for a qualitative understanding of certain physical processes in these compounds, the results of LDA+DMFT calculations can be directly compared to experimental measurements. Spectroscopic measurements contain a huge amount of information. As we have seen in the discussion at the end of the previous chapter many of the different techniques are complementary and often only a combined analysis can yield a conclusive picture of the system of interest.

This chapter is devoted to the actual application of the LDA+DMFT formalism for the comparison between calculations and experiment for real compounds. More precisely, most of the systems which will be discussed in this work belong to the family of the transition–metal oxides – maybe the most prominent class of strongly correlated electron systems.

The first one to be presented is the well known vanadium sesquioxide V_2O_3 for which photoemission, optical conductivity, X–ray absorption data, and its implications for the doping-, pressure-, temperature phase diagram will be discussed in detail. In this respect the X–ray absorption data are special since we employ a combination of the LDA+DMFT and the briefly mentioned CI calculations in order to interpret experimental results. This interpretation, together with new measurements (performed by L. Baldassare et al.) and calculations of optical conductivity sheds new light on the still controversial issue of the metal–to–insulator transition in the V_2O_3 compound. The second system is the transition–metal chalkogenide $NiSe_xS_{2-x}$. We will discuss calculations of the photoemission spectra and it will become clear, that simple models with d–states only are not really applicable which motivates the extension of the LDA+DMFT basis–set.

3.1 Optics and X–ray absorption of V_2O_3

Some materials present metal–to–insulator transitions without any changes in crystal structure or long-range magnetic order. These phenomena, known as Mott–Hubbard transitions, constitute a hallmark of strong electronic correlations. The physics emerging in the vicinity of these transitions is highly non–trivial and the properties of such materials depend crucially on small changes in the electronic structure induced by external parameters [89, 44]. The isostructural MIT in Cr-doped V_2O_3 is considered as the textbook example of a Mott transition, which occurs between a paramagnetic insulator (PI) and a paramagnetic metallic (PM) phase by changing doping level (x), temperature (T) or pressure (P) [124]. Several features of the MIT have been successfully clarified by resorting to realistic many–body calculations [68]. Yet, contrary to common assumptions, a growing number of experimental facts are revealing that the MIT and the resulting phases are also strongly dependent on the route followed through the transition. Specifically, a "common wisdom" was established, as we sketch it in Fig. 3.1, that the appliance of pressure on insulating Cr–doped V_2O_3 could recover the undoped metallic phase (even an empirical relation of external pressure and doping has been proposed [124]). However, as a result of our work we will see, that the metal–

3.1 Optics and X–ray absorption of V_2O_3

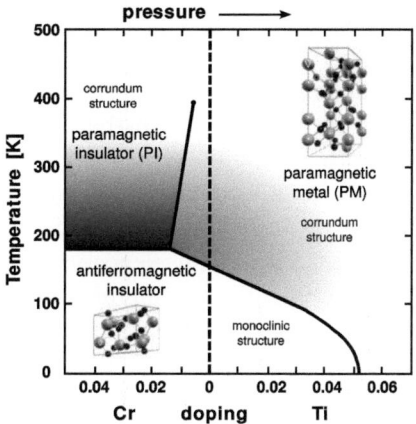

Fig. 3.1: Phasediagramm of V_2O_3 in the temperature vs. doping plot displaying three phases: paramagnetic metallic (PM), paramagnetic insulating (PI), and antiferromagnetic insulating (AF). In the PM and PI phase the compound crystallizes in the corundum structure, whereas the low temperature AF phase shows a monoclinic lattice structure. The monoclinic and corundum structure unit cells are shown as small insets.

lic phases of the undoped sample and the one of the doped compound under pressure are actually quite different concerning their electronic structure.

3.1.1 The story so far

At first, let us summarize the basic facts and review some of the theoretical work that has been put forward, thereby also defining the necessary terms for our own analysis. Starting from the phase diagram and the crystal structure, we will briefly recall the work that has been done, including previous LDA+DMFT studies.

phase diagram and crystal structure In Fig. 3.1 we show the phase diagram of V_2O_3 spanned in the temperature–doping space displaying the above mentioned three phases (PM,PI,AF) of the compound. At ambient conditions V_2O_3 is a metal and crystallizes in the corundum structure with four vanadium atoms in the primitive unit cell. The conventional and the primitive unit cells are sketched in Fig. 3.2 and it can be seen, that respectively two

3 Bulk 3d–transition metal compounds: Theory vs. Experiment

vanadium atoms form "pairs" which are oriented along the crystallographic c–axis. Upon cooling below 150K, a peculiar antiferromagnetic order sets in and the system becomes insulating, accompanied by a monoclinic structural distortion. On the other hand, the system can be tuned by doping with chromium or titanium or the application of external pressure. Usually it is assumed, that doping and pressure can be seen as equivalent routes through the phase diagram. Within this chapter, however, the pressure/doping equivalence scheme will be shown to be inconsistent with recent experimental measurements of the optical conductivity and x–ray absorption. Above the Néel, temperature, the corundum crystal structure does not change as a function of pressure or doping. However, upon Cr doping a first order isostructural metal–to–insulator (PM–to–PI) transition takes place (see Fig. 3.1) which evoked several theoretical attempts to describe this MIT as a genuine Mott–Hubbard transition. While the MIT is associated to changes in the lattice structure and the atomic positions [126, 167], it is important to notice that x–ray diffraction showed that for a given temperature the structure within one phase *does not change upon doping* [167]. It was later also observed by Park *et al.* with vanadium L–edge x–ray absorption spectroscopy that this holds also for the electronic ground state of the system (see Table 1 of [153]). Therefore we shall adapt the nomenclature of Robinson [167] and refer to the lattice structure of the metallic and insulating phase *at ambient pressure* as α– and β–phase respectively.

The electronic configuration of atomic vanadium is $[Ar]3d^34s^2$, which means, that in the three–valent oxidation state we find a $3d^2$ configuration realized. In the corundum type structure the vanadium atoms are coordinated by oxygen ligands in a trigonally distorted octahedral fashion (left hand side of Fig. 3.2). Hence, the cubic part of the crystal field splits the d–states into the lower lying t_{2g} and the higher lying e_g states. The trigonal distortion[1] acts like a compression along one of the three–fold axes of the octahedron

---
[1] Let us, already in this introduction, remark that the actual crystal field breaks one significant point symmetry on the vanadium sites, namely inversion in the c–direction. This is related to the different distances of the neighboring vanadium atoms along the c–axis. While this effect is negligible for most of the discussion, it will be of great importance for the selection rules of the polarization dependent XAS results later on.

3.1 Optics and X-ray absorption of V_2O_3

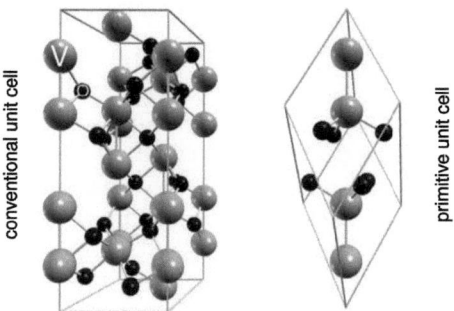

Fig. 3.2: Conventional (left hand side) and primitive (right hand side) unit cell of V_2O_3 in the corundum structure. The coordination polyhedron of the vanadium atom is an octahedron of oxygen ligands distorted along the C^3 axis (i.e. two opposing planes of the octahedron are "squeezed" together). The point group of a vanadium site is D_{3d}. Moreover, inversion symmetry along the closest V–V bond is broken which leads to an onsite mixing of V 3d and V 4p states.

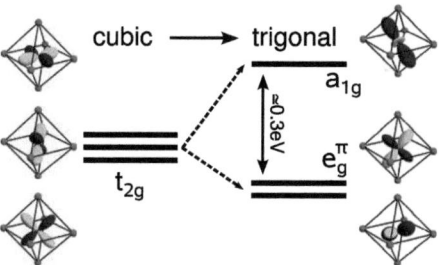

Fig. 3.3: Level splitting of the vanadium t_{2g}-states: due to the trigonal distortion the t_{2g}-states split up in an e_g^π doublet and an a_{1g} singlet. From LDA results we estimate approximately 0.3 eV (in the plotted direction) for this splitting. The plotted orbitals are spherical harmonic functions which display the symmetry of the states. These plots can be nicely compared to the NMTO Wannier function plots of Saha–Dasgupta et al. [178]

47

(i.e. squeezing two opposite sides together). As a result the degeneracy of the lower lying t_{2g} states is lifted and they are split into a single a_{1g} and the twofold degenerate e_g^π states. This level splitting, together with a plot of the respective angular part of the (atomic) wave function, is sketched in Fig. 3.3. To indicate the difference to the t_{2g} states, the higher lying cubic e_g states (which are not split by the trigonal distortion) get an additional index e_g^σ in order to distinguish them from the e_g^π. The σ accounts for their orientation towards the ligands, with which they form σ bonds.

Since the e_g^σ are pushed up in energy by the crystal field, the two vanadium d–electrons populate the t_{2g} levels. One of the crucial aspects concerning the understanding of the MIT is the specific occupation of these t_{2g} states. In an early work, Castellani et al. [28, 27, 29] assumed a strong hybridization of the V–V pairs oriented parallel to the rombohedral c–axis resulting in a strong bonding and antibonding splitting of the a_{1g} states. In this case, with the bonding states filled there would be one electron remaining in the twofold degenerate e_g^π states and the compound could be described by a quarter filled $S = 1/2$ Hubbard model. However, later experimental evidence demonstrated [133, 152, 50, 153] that the ground state of the system is more complicated and should rather be described as a $S = 1$ state consisting of a mixture of a_{1g} and e_g^π.

Moreover, it is precisely the coefficients in the linear combination of a_{1g} and e_g^π for the ground state which allow for a quantitative distinction of the PM, PI, and AF phases. The XAS vanadium L–edge study of Park et al. explored the phase diagram by means of temperature and doping and summarized the respective ratios of a_{1g} and e_g^π in table 1 of [153]. As we mentioned earlier, their results turned out to be consistent with the x-ray diffraction data for the lattice of Robinson [167], and showed that, within the PM α– and the PI β– phase, there is no change in the ground state composition for different doping. One of our main results, which will be discussed later in this chapter, is, that this is not true for the pressurized metallic phase.

ab initio LDA calculations First *ab initio* LDA band structure calculations for V_2O_3 were performed by Mattheiss [122]. Not surprisingly the results

3.1 Optics and X-ray absorption of V_2O_3

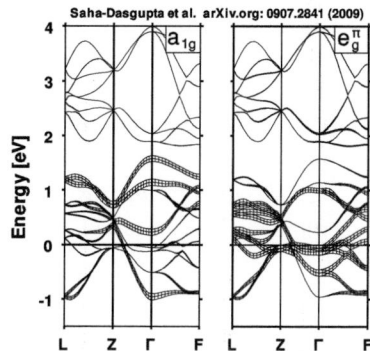

Fig. 3.4: V_2O_3 bandstructure from LDA with "fat bands" indicating the orbital character (plot taken from Saha–Dasgupta et al.[178])

neither captured the insulating character of the Cr doped PI phase nor the signatures of the strongly correlated character of the undoped PM phase (for example the photoemission spectral weight identified with the lower Hubbard band). Yet, even on the LDA level, representing the starting point also for the analysis discussed in this work, some valuable information can be obtained. In Fig. 3.4 a plot of the LDA band structure is shown in the rombohedral representation taken from Saha–Dasgupta et al.[178] with $\varepsilon_F = 0$. In the two panels the respective a_{1g} and e_g^π character is indicated by the width of the lines by means of the so called "fat band" representation.

First of all it can be seen that the t_{2g} part is nicely separated from the rest of the bands: Genuine oxygen p–bands are lower lying than the displayed energy range and the e_g^σ can be identified as the bands at ≈ 2 eV to 4 eV. In principle it is allowed by symmetry in the trigonal environment that the e_g^π states mix with the e_g^σ states. Yet, from the fat bands we find this mixing to be rather small.

Let us now turn to the a_{1g} bands (left panel Fig. 3.4). The previously mentioned bonding–antibonding splitting due to the V–V pair can be seen at the Γ–point to be > 2 eV, where the a_{1g}–character is pure. The strongest dispersion we observe along Γ–Z where the main contribution stems from a_{1g}–a_{1g} hopping. However, the dispersion of the a_{1g}–bands along the other directions is clearly not small which is a consequence of a_{1g}–e_g^π hybridization. While the a_{1g} and e_g^π states are orthogonal eigenstates *locally* in a trigonal crystal field they still hybridize in a *non–local* way, i.e., intersite a_{1g}–e_g^π hopping also in the ab–plane. As it was remarked by Elfimof et al. [52] these kinds of hopping are important for the shape of the a_{1g} states. With the help of the fat bands in Fig. 3.4, we compare the a_{1g} and e_g^π–character: We clearly see that only at the high symmetry points in the Brillouin zone the non–local a_{1g}–e_g^π hybridization is zero. In fact, this result contradicts also theoretically the validity of quarter filled e_g^π states and the $S = 1/2$ scenario.

previous LDA+DMFT studies Starting from the LDA results, first LDA+DMFT calculations were performed and compared to photoemission and XAS experiments by Held et al. [82] and Keller et al. [102]. However, these first studies were approximative in the sense that they used the local projected density of states instead of a t_{2g} Wannier functions Hamiltonian[2]. In 2007 Poteryaev et al. [157] performed new LDA+DMFT calculations employing such a downfolded NMTO t_{2g} Wannier functions Hamiltonian provided by Saha–Dasgupta et al. [178]. In this work the authors discuss in detail the changes in the effective quasiparticle band structure caused by the correlations and the corresponding self–energies. Further, they also give a comparison of the LDA+DMFT spectral functions to experimental photoemission data. The Wannier Hamiltonian approach (as opposed to the projected DOS method) yielded an improved agreement with experiment compared to earlier works and the analysis of the self–energy showed a strong orbital dependence of the coherence on the orbital character. Following the work of Poteryaev et al. [157], Tomczak [201] and Tomczak and Biermann [202] discussed the optical conductivity of the compound introducing corrections for the calculation of the Fermi–velocities associated with the non–monoatomic basis of V_2O_3 – an important issue also for this work which will be discussed in the next section. The last work that should be mentioned is the joint experimental/theory paper by Baldassare et al. [14] in which the authors show that the slight change of the lattice parameters due to temperature, drive the system towards the paramagnetic–insulating (PI) state. Their results underline,

---

[2]this approximation becomes exact in the case of degenerate bands and does not capture correlation mediated level shifts, i.e., "crystal field enhancement"

how sensitive strongly correlated systems are with respect to the change of external parameter – even more so in the vicinity of a correlation driven Mott transition.

The key interest of our work, which will be discussed in the following, is to shed new light on the actual ground state of V_2O_3 at different points in the phase diagram Fig. 3.1. Special attention is paid to the insulating and metallic phase of the 1.1% Cr–doped sample in the vicinity of the MIT as well as to the comparison between the metallic phase of the undoped sample at ambient conditions and the Cr–doped sample under external pressure.

3.1.2 The low energy t_{2g} NMTO model

The first step of the LDA+DMFT calculations is the derivation of Hamiltonian (2.26) after the bandstructure calculation via NMTO downfolding or Wannier projections. The Hamiltonian is constructed to capture the relevant degrees of freedom of the system for low energy scales on a reduced basis set. In the case of V_2O_3 we used a model obtained by the NMTO method and the full LDA Hamiltonian was downfolded on the t_{2g} sub–space around the Fermi energy. As described above (see Fig. 3.3), the t_{2g} states are decomposed into a single a_{1g} and two degenerate e_g^π states. However, if we look closely at the bandstructure in Fig. 3.4 we find twelve t_{2g} bands instead of three. The reason for this is simply that there are four vanadium atoms in the primitive unit cell which means, that we obtain a 12 by 12 Hamiltonian as a function of k for V_2O_3 from the downfolding. (For a detailed discussion of the downfolding procedure of the V_2O_3 model see Saha–Dasgupta *et al.* [178]). Yet, although the LDA Hamiltonian is a twelve–band dispersion matrix, the actual DMFT calculation can be performed with no more effort than a three band calculation. The reason for this is simply that all four vanadium atoms in the unit cell are located on equivalent sites which means that they are related to one another by symmetry transformations. In other words, each of the four vanadium atoms experiences the same environment and, hence, has

3 Bulk 3d–transition metal compounds: Theory vs. Experiment

the same *local* eigenstates. As a consequence, the k–integrated local Green function in (2.24) can be written in a basis in which we obtain four equal diagonal blocks with respect to the site index. As a consequence, the orbital labels a_{1g} and e_g^π are good quantum numbers *locally*. Such a local basis set is a necessary condition for the formulation of the local interaction parameter U and a correct definition of the *local* DMFT self energy.

We explicitly write the local Green function as:

$$\underline{\underline{G}}^{\text{loc.}}(\omega) = \frac{1}{V_{\text{BZ}}} \int_{\text{BZ}} d^3 k \frac{1}{(\omega + \mu)\underline{\underline{1}} - \underline{\underline{\epsilon}}^{\text{LDA}}(\mathbf{k}) - \underline{\underline{\Sigma}}(\omega)}$$

$$= \begin{pmatrix} \underline{\underline{G}}^{\text{I}} & & \underline{\underline{G}}_{\text{hyb.}} \neq \underline{\underline{0}} \\ & \ddots & \\ \underline{\underline{G}}_{\text{hyb.}} \neq \underline{\underline{0}} & & \underline{\underline{G}}^{\text{IV}} \end{pmatrix} \quad (3.1)$$

where the roman numerals serve as a site index, and, as mentioned

$$\underline{\underline{G}}^{\text{i}} = \begin{pmatrix} G^{\text{loc.}}_{a_{1g}} & 0 & 0 \\ 0 & G^{\text{loc.}}_{e_g^\pi} & 0 \\ 0 & 0 & G^{\text{loc.}}_{e_g^\pi} \end{pmatrix} ; \quad i = \text{I}, \cdots, \text{IV} \quad (3.2)$$

the diagonal blocks are equal for each site i. Hence, in order to calculate the local DMFT self energy, which clearly has to be the same for all four (locally equivalent) sites, we have to "cut out" (i.e. project out) only the first diagonal block after the k–integration, and proceed with the calculation of the DMFT self energy in the normal way. The resulting self energy is a diagonal 3 by 3 matrix $\underline{\underline{\Sigma}}^{\text{I}}(\omega)$ and is used in the next iteration to construct the full 12 by 12 diagonal matrix $\underline{\underline{\Sigma}}(\omega)$ taking the equality of the four vanadium sites into account

$$\underline{\underline{\Sigma}}(\omega) = \begin{pmatrix} \underline{\underline{\Sigma}}^{\text{I}}(\omega) & 0 & 0 & 0 \\ 0 & \underline{\underline{\Sigma}}^{\text{I}}(\omega) & 0 & 0 \\ 0 & 0 & \underline{\underline{\Sigma}}^{\text{I}}(\omega) & 0 \\ 0 & 0 & 0 & \underline{\underline{\Sigma}}^{\text{I}}(\omega) \end{pmatrix} \quad (3.3)$$

This full self energy then enters equation (3.1) for the calculation of the next local Green function. In this respect V_2O_3 is by no means an exceptional case. In all compounds studied in this thesis with more than one (correlated) d– or f– atom in the unit cell this situation occurs. Further, it is important to strongly stress at this point that we do not make additional approximations with the procedure described above. From the DMFT point of view, i.e., the local perspective, the V_2O_3 calculation *is* just a three orbital problem. However, the situation will of course change dramatically if the sites of the atoms differ locally. Such cases will be discussed later in chapter 5 where we present conceptual extensions of the standard single site DMFT calculations. Besides the additional step of projecting out the local part of the full Green function the LDA+DMFT calculation of V_2O_3 is straight forward as we discussed in chapter 2. Let us therefore turn to the spectroscopic data and our theoretical interpretation.

3.1.3 Optical conductivity: Phase separation around the MIT

The first work we will discuss, are measurements of infra red optical conductivity carried out in the group of Prof. S. Lupi in the University "La Sapienza" in Rome (Italy). This work has a twofold goal: on the one hand, to clarify the behavior of the 1.1% Cr–doped compound around the metal to insulator transition, and on the other hand, to perform an experimental check of the pressure doping equivalence. The motivation of the former analysis is the following: In the past much effort has been put into the understanding of the transition between the PM and the PI phase Fig. 3.1. However, somehow less, or at least less concrete, attention was paid to the local strain that occurs in the lattice in the Cr–doped compounds [126, 65], even though, for $(V_{0.989}Cr_{0.011})_2O_3$, the presence of a structural phase separation, between the PM α– and PI β–phase, by the Cr–doping has been clearly stated [126, 65, 167]. Other experimental studies also support the idea that the Cr– atoms in $(V_{0.989}Cr_{0.011})_2O_3$ could play the role of β–phase "condensation nuclei":

Resistivity measurements, for example, show that the conducting phase of weakly Cr–doped samples shows a bad metallic behavior, differently from the undoped compound [108]. Moreover, so called extended x–ray absorption fine–structure spectroscopy (EXAFS) measurements showed that the presence of Cr contracts the Cr–V bonds, inducing a concomitant elongation of V–V pair bonds [65]. Such "long" V–V pair bond is associated to the β PI phase [125], as shown also by theoretical calculation using LDA+DMFT [82]. Therefore it may be hypothesized that, within an insulating matrix host, metallic–like "islands" are formed around the Cr impurities [42, 166]. On this basis, the PM–PI MIT has been suggested to have also a percollative nature [65]. Pressure–dependent transport studies by Limelette et al. [114] were also used to show that across the PI–PM first order transition a large hysteresis occurs. This points to a non–trivial role of the lattice and its distortions due to the Cr doping, which has however been almost disregarded, or drastically simplified when defining the "standard" phase diagram. This has been established by means of resistivity data only and suggests the equivalence of doping and pressure.

The relation between such hysteresis and the above mentioned coexistence of α and β–phases has not been hitherto clarified. It is these unknowns at which our investigation aims.

Experimental results We report the experimental results of the Rome group in Fig. 3.5: In the upper panel the positions in the phase diagram where the spectra were taken are marked. The spectra are plotted in the lower panel in the corresponding color: on the left hand side several spectra for different temperatures are shown together with their values for the DC conductivity σ_{DC} at $\omega \to 0$, whereas on the right hand side the T=200K spectra for the undoped and the 1.1% Cr–doped samples are compared.

Let us start with the discussion of the temperature dependent data. Shown in Fig. 3.5 (lower panel left hand side) are $(V_{0.989}Cr_{0.011})_2O_3$–spectra in the temperature range between 500K and 200K (the sharp peaks around 500cm^{-1} cor-

3.1 Optics and X-ray absorption of V_2O_3

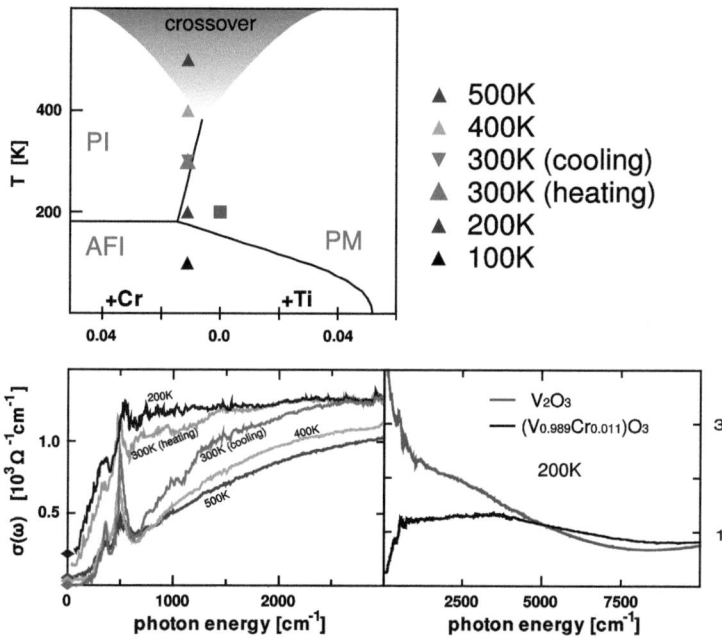

Fig. 3.5: IR optical conductivity for V_2O_3. Upper panel: phase diagram with marks at the positions where the spectra were taken. Lower panel left hand side: spectra for the 1.1% Cr–doped $(V_{0.989}Cr_{0.011})_2O_3$ sample at different temperatures ranging from 500K down to 200K. At 300K, hysteresis can be found and the spectra differ severely depending from which phase the point is approached. Lower panel right hand side: comparison between the $(V_{0.989}Cr_{0.011})_2O_3$ spectrum and the spectrum of undoped V_2O_3 at 200K.

respond to phonon resonances and are of no further interest for the present discussion). Our starting point is in the cross–over region of the transition at 500K. Cooling down we obtain the spectrum at 400K and at 300K. At 300K, however, we are in the direct vicinity of the transition line. Hence, we can, in fact, find a qualitatively different (PM–phase) spectrum at 300K if we approach the same point by heating up from lower temperatures. That is, we observe hysteresis. The last spectrum we show in the plot was taken in the PM phase at 200K. The first three spectra, from 500K down to 300K (upon cooling) display the gapped shape which we expected as the hallmark of the insulating nature of the PI β–phase. The spectra show no Drude peak and, only at elevated temperatures, gain minimal spectral wait at $\omega \to 0$. The remarkable, and far from trivial, spectra are the ones in the PM phase at 300K (upon heating) and at 200K. We recall that already resistivity measurements have shown a bad metallic behavior for the Cr–doped sample as opposed to the undoped compound. Yet, how dramatic the difference to the undoped sample really is, can be seen clearly in Fig. 3.5 (lower panel right hand side) where we show a direct comparison of the $(V_{0.989}Cr_{0.011})_2O_3$–spectrum at 200K and the spectrum of the undoped sample at the same temperature. Whereas the spectrum of the undoped sample shows the behavior that is expected from a metallic phase, including a well pronounced Drude peak, the shape of the spectrum of the Cr–doped compound is rather unexpected: It does not show a Drude peak, but neither has it a gap like in the insulating regime as the spectral weight around $\omega \to 0$ is non–negligible. This fact is a clear support for a scenario of an inhomogeneous (i.e. α–β mixed) metallic phase. On the contrary, when comparing the behavior of $(V_{0.989}Cr_{0.011})_2O_3$ and $(V_{0.972}Cr_{0.028})_2O_3$ within the PI phase, only small differences appear (not shown here).

The new interesting experimental facts are motivation enough for us to revisit the compound again with the help of LDA+DMFT in order to understand the features that are displayed more fundamentally. Performing this analysis we want to test the hypothesis of the mixed α–β phase scenario.

3.1 Optics and X-ray absorption of V_2O_3

LDA+DMFT analysis The starting point for our theoretical LDA+DMFT analysis is the downfolded NMTO Hamiltonian described in the previous section for the α– and the β–phase respectively. In our DMFT code, we employ the Hirsch Fye quantum Monte Carlo method. The calculations were carried out at an inverse temperature of $\beta = 20$ eV$^{-1} \approx 580$K and with interaction parameters $U = 4.0$ eV and $J = 0.7$ eV. After convergence of the DMFT self consistent loop the single particle Green function on the imaginary time τ axis has been analytically continued by means of the Maximum Entropy Method (see chapter 2). Next, we extracted the local self energy on the real axis in order to calculate the optical conductivity measured in the experiment. In the following we will first discuss the direct results, i.e., spectral functions and local self energy thereby also comparing them to the previous data from Poteryaev et al. [157]. Afterwards we present the calculation of the optical conductivity.

LDA+DMFT results for ($V_{0.989}Cr_{0.011}$)$_2O_3$ In Fig. 3.6 we report the orbital–resolved spectral function (upper panel) as well as the according self energies (lower panels). In the plots we set the Fermi energy to $\varepsilon_F = 0$ and plot the sum of the two degenerate e_g^π spectra in dark gray, the a_{1g} spectrum in light gray, and the total spectrum, i.e., the sum over all, in black. We summarize the quantities for the α– and the β–phase on the left hand and right hand side of the panels respectively. Overall our results agree with the results of the previous LDA+DMFT analysis by Poteryaev et al. [157], although we performed the calculations at slightly lower $U = 4.0$ eV values (in [157] $U = 4.2$ eV)[3]. The self energies, both real and imaginary parts, display a strongly orbital dependent character. The real part acts like an orbital dependent renormalization of the chemical potential or, in other words, as it is called in [157] as an "effective crystal field"[4] whereas the imaginary part is a measure of lifetime/coherency of the excitations in the associated band. However, the self energy depends

---
[3] The reason for our choice is the sensitive dependence of the optical gap on this parameter.
[4] This expression should be used with care since the term "crystal–field" as it was originally introduced (and as it also will be used later) in the section about the CI calculations for the XAS has a different origin, namely the electrostatic and covalent interaction with the ligands

3 Bulk 3d–transition metal compounds: Theory vs. Experiment

INFO: Interaction parameters for V_2O_3

From the technical perspective, we need to elaborate more detailed on the important issue of choosing the appropriate values for interaction parameters of a specific compound and the theoretical method that is employed. V_2O_3 presents a good example in that respect, since in the literature several different values for U and J can be found. The confusion about these parameters partly stems from the improvement in the estimates of their values over the time and partly from the differences in the numerical techniques. A constrained LDA calculation (for the monoclinic antiferromagnetic phase) by I. Solovyev et al. [189] yields the values of $U = 2.8$ eV and $J = 0.93$ eV – parameters that later on were employed in some works [56, 83, 82]. Yet, constrained LDA gives unfortunately only very rough estimates of the values for U, which not only crucially depend on the electronic structures, but also on the basis set of the model at hand because it is highly sensitive to screening. For example, U has to be chosen much lower in the case of a LDA+U calculation (for V_2O_3 $U = 2.8$ eV) in comparison to DMFT values ($U \approx 4$ eV) in order to overcome the deficiency of the static mean field nature of LDA+U which overestimates ordering and gaps (see e.g. Sangiovanni et al. [181]). We choose the parameter $U = 4.0$ eV following the philosophy of Held et al. [83] and Poteryaev et al.[157] that the value of U should be consistent with the correct physics of V_2O_3, i.e., the MIT is reproduced within LDA+DMFT. Therefore, it is not surprising that our choice of U agrees well with the LDA+DMFT study of Poteryaev et al. since they employ the same up to date LDA+DMFT scheme as we do. Let us also remark here that we have already studied separately this specific issue [205]. The result of our analysis clearly demonstrates that the appropriate U for the LDA+DMFT calculation should be chosen in the range 4.0 eV $< U <$ 4.2 eV, as we did in the present calculations. Considerably smaller and larger values of U would either lead to the disappearance or a huge overestimation of the spectral, and as to be seen also optical, gap in the PI phase.

3.1 Optics and X-ray absorption of V_2O_3

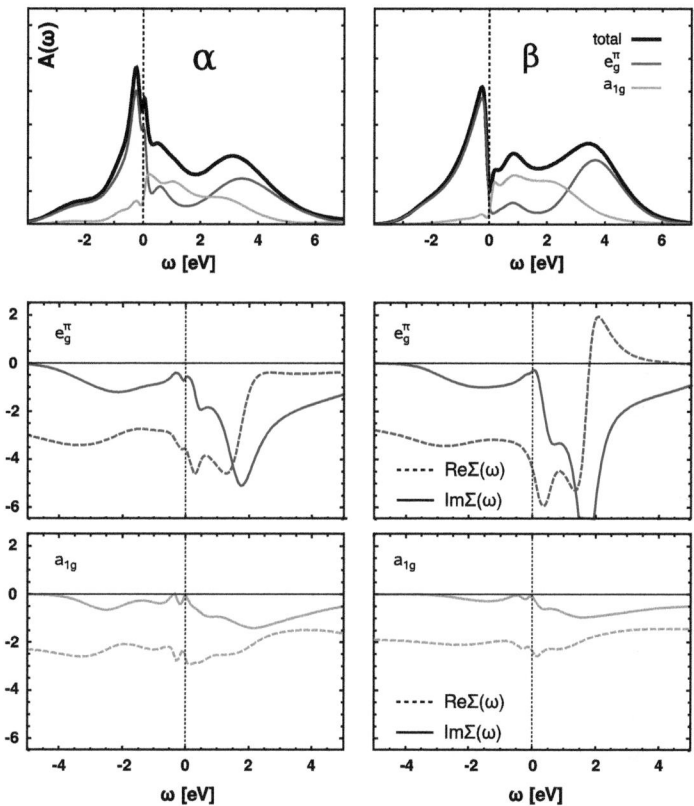

Fig. 3.6: LDA+DMFT results for $(V_{0.989}Cr_{0.011})_2O_3$ in the α phase (left hand side) and the β phase (right hand side) for $U = 4.0$ eV. In the upper panels we show the LDA+DMFT spectral functions resolved in orbital labels and coded by their color (see legend). The spectra show metallic behavior (coherent excitations around the Fermi energy) for the α phase and gapped insulating behavior for the β phase. In the lower panels we show the self energies for the e_g^π (middle) and a_{1g} (bottom) states.

3 Bulk 3d–transition metal compounds: Theory vs. Experiment

also on the filling of the respective orbitals and in a hybridized system like the t_{2g} states of V_2O_3 it is a very involved quantity: Although the self energy is diagonal we see from equation (3.1) that its connection to the Green function, and hence the spectrum, involves an inversion so that the orbitally resolved information is, in a way, convoluted.

The spectral functions for the α and β–phase are quite similar, except for the strongly renormalized coherent quasiparticle excitations of the correlated metallic α–phase around the Fermi energy. Of course, the differences are expected to be sharpened up at lower temperatures. From the orbital–resolved spectra we can obtain valuable insight. Let us have a closer look at the incoherent part of the spectrum, i.e., the Hubbard bands. The basic features can be understood as follows: As it was discussed in previous works (e.g. [83]) and also will be confirmed later by our XAS study the predominant local configuration on the V atoms has two spin–aligned electrons in the e_g^π orbitals, i.e., a $|e_g^\pi e_g^\pi\rangle$ spin–1 configuration, with some admixture of $|a_{1g} e_g^\pi\rangle$ spin–1 configurations. For a simple picture let us first consider the lower Hubbard band (LHB), that is, the electron removal part of the spectrum. We recall the relevant onsite interaction parameters to be the intra–orbital interaction U, the inter–orbital interaction V, and the spin–coupling constant J. Furthermore, in cubic (or close to cubic) symmetry the relation $V = U - 2J$ holds. Starting either from the $|e_g^\pi e_g^\pi\rangle$ or the $|a_{1g} e_g^\pi\rangle$ configuration, the removal of an electron will result in an energy gain of $V - J$ (≈ 1.9 eV in our case) which is in agreement with the position of the LHB. The only structure, i.e. splitting, which occurs is the crystal field potential differences of the e_g^π and a_{1g} states. This energy scale, however, is below the resolution of our spectra at high $|\omega|$. For the upper Hubbard band, i.e., the electron addition part, the situation turns out to be a little bit different. The additional electron can either populate an e_g^π or an a_{1g} state. Then the process $|e_g^\pi e_g^\pi\rangle \rightarrow |e_g^\pi e_g^\pi e_g^\pi\rangle$ or $|a_{1g} e_g^\pi\rangle \rightarrow |a_{1g} a_{1g} e_g^\pi\rangle$ will cost an energy of $U + V$. The additional electron interacts via U with one of the other two electrons, and via V with the other one. Transitions $|e_g^\pi e_g^\pi\rangle \rightarrow |e_g^\pi e_g^\pi a_{1g}\rangle$ or $|a_{1g} e_g^\pi\rangle \rightarrow |e_g^\pi e_g^\pi a_{1g}\rangle$ only cost $2V$ or $2V - 2J$ depending on the respective spin alignment. Consequently, the UHB is split into two main features which we can find around 1 eV and $4 - 5$ eV. We conclude that i) the split of the UHB depends apparently strongly on the choice of J and ii)

3.1 Optics and X-ray absorption of V_2O_3

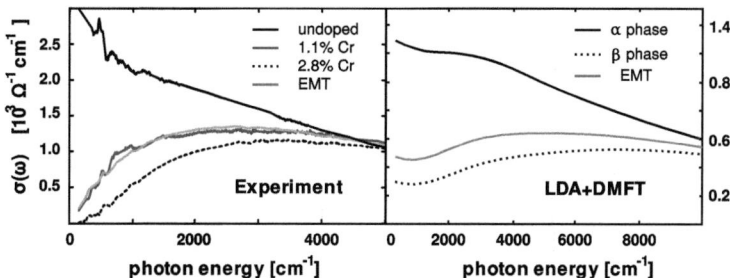

Fig. 3.7: Phase mixing: Comparison of experimentally observed (left hand side) and LDA+DMFT calculated (right hand side) optical conductivity. In black we plot the spectrum for the α phase, for which we take the experimental spectrum of the undoped sample. The β phase, for which we take the experimental spectrum of the 2.8% Cr doped sample is shown in dashed black. The measured spectrum of the 1.1% Cr doped sample (dark gray) can be fitted by a mixture of α and β phase spectra (light gray) within the *effective medium approximation*

this split is and responsible for the small width of the gap compared to the interaction parameters[5].

Let us, finally, turn to the optical conductivity. The calculation of the optical conductivity has been performed according to section 2.4. It should be remarked, that the non–monoatomic basis of the crystal leads to corrections in the calculation of the Fermi velocities even in the Peierls approximation as it is discussed by Tomczak [201] and Tomczak and Biermann [202].
In Fig. 3.7 we show a comparison of experimental data (left hand side) and LDA+DMFT data (right hand side). We calculated the optical conductivity also for both α and β–phase. The LDA+DMFT optical conductivity of the β–phase (right hand side: dashed) shows a gapped behavior, as it is expected for the PI phase. The fact that it does not extrapolate to zero at energies lower than 1000cm^{-1} is due to the temperature of $\beta^{-1} = 0.2\text{eV} \approx 500$ K assumed for the DMFT(QMC) calculations. Further, when we compare it to the experimental

---
[5]This observation explains also why the attempt to handle the gap (actually the optical gap) with a one band Hubbard model [172] led to unphysically small values for the interaction parameter.

data of the 2.8% Cr–doped sample, deep in the PI phase, we see that the gap of our calculated σ_β is a little bit too large. The reason for this overestimation is an extreme sensitivity of the calculation on the choice of U and J as it was already mentioned before[6]. The calculated α–phase optical conductivity (right hand side: black) shows an overall good agreement with the experimental data taken for the undoped compound (left hand side: black). At $\omega \to 0$ we can distinguish the typical Drude peak contribution of the PM phase.

The most interesting spectrum, however, corresponds to the experimental data taken for the 1.1%Cr–doped sample at 200K (left hand side: dark gray): As mentioned above this spectrum is strange in its shape (with neither Drude peak nor gap) and belongs to a state that is, according to the resistivity measurements, a bad metal. The discrepancy between the idea that the PM phase can be seen as a uniform metallic phase and the experimental evidence is further enhanced by our LDA+DMFT calculations. Specifically, as the lattice parameters practically do not change within the α–phase the LDA+DMFT spectrum of the 1.1%Cr–doped sample at 200K and of the undoped compound are basically indistinguishable. Hence, our calculations support the hypothesis of an α–β phase mixture in the bad metal region. To test this hypothesis further we resort to a semi–empirical formula of the *effective medium theory* (EMT) [42, 25] which provides a simple way of approximating the dielectric constants $\bar{\varepsilon}(\omega) = (1 - \sigma(\omega))$ for mixtures of insulating and metallic phases. Within the EMT the effective constant $\bar{\varepsilon}(\omega)_{\text{eff.}}$ satisfies the condition

$$f\frac{\bar{\varepsilon}_{\text{met.}}(\omega) - \bar{\varepsilon}_{\text{eff.}}(\omega)}{\bar{\varepsilon}_{\text{met.}}(\omega) + \frac{1-q}{q}\bar{\varepsilon}_{\text{eff.}}(\omega)} + (1-f)\frac{\bar{\varepsilon}_{\text{Ins.}}(\omega) - \bar{\varepsilon}_{\text{eff.}}(\omega)}{\bar{\varepsilon}_{\text{Ins.}}(\omega) + \frac{1-q}{q}\bar{\varepsilon}_{\text{eff.}}(\omega)} = 0 \qquad (3.4)$$

Where f and q are free fitting parameters which are phenomenologically related to the size and relative densities of the "islands" of the two constituent phases. For further information about this approach we also refer to [164]. Now we take the optical conductivity spectra measured in the undoped sample and the 2.8% Cr–doped sample as the α– and β–phase spectra respectively and use Eq. (3.4) to fit the experimental spectra of the 1.1% Cr–doped

---

[6] A slightly larger U, like it was used, e.g., in [157] would result in an even larger gap.

3.1 Optics and X–ray absorption of V_2O_3

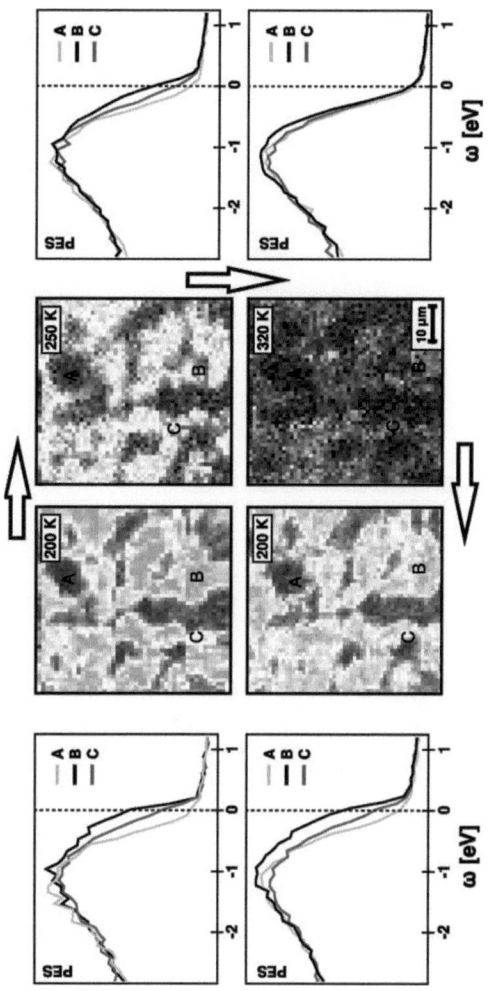

Fig. 3.8: Representative intensity ratio images and corresponding photoemission spectra at different temperatures over a 50 μm by 50 μm area obtained by scanning photoemission microscopy using photons at 27 eV. Inhomogeneous properties are found within the PM phase $T = 220$ K and 260 K, where metallic (light) and insulating (dark) domains coexist. A homogeneous insulating state is instead obtained in the PI phase at 320 K. After a whole thermal cycle the structure of the inhomogeneous distribution is recovered, indicating the presence of stable "condensation nuclei".

3 Bulk 3d–transition metal compounds: Theory vs. Experiment

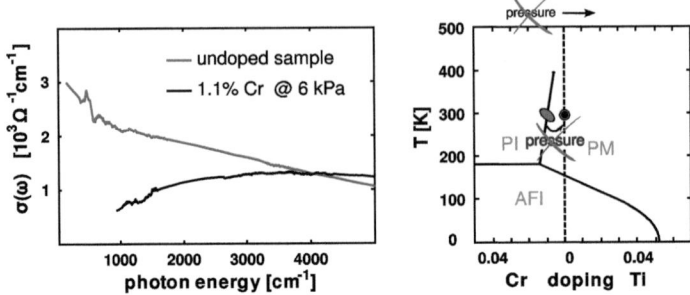

Fig. 3.9: Experimental measurements of the optical conductivity for the undoped sample (light gray) and the 1.1% Cr doped sample at 6 kPa. For this pressure the "common wisdom" assumed a full recovered PM phase equivalent to the undoped compound PM phase. The experimental data clearly proofs this assumption to be incorrect and suggests to abandon the concept of pressure equals inverse doping.

sample. For the values $f = 0.42$ and $q = 0.35$ an excellent agreement can be found which is plotted in Fig. 3.7 (left hand side: compare light and dark gray). From a theoretical point of view, it is even possible to directly use the LDA+DMFT spectra for the α– and β–phase of $(V_{0.989}Cr_{0.011})_2O_3$ as an input for the EMT. Also in this case, with the same values of f and q we obtain a satisfying agreement with the experimental data.

To sum up, the experimental measurements of the optical conductivity together with the theoretical interpretation by means of LDA+DMFT strongly support the scenario of a mixed phase state for the 1.1% Cr–doped compound at 200K. It will be seen in the next section that the complementary x–ray absorption spectroscopy also speaks for this result. Recently, also spatially resolved photoemission microscopy data was obtained for the 1.1% Cr–doped compound in the group of Prof. M. Marsi Université Paris–Sud. The measured spectra spectacularly enforces our interpretation: In Fig. 3.8 we show the microscopic images together with the PES for the labeled positions. At 200K one clearly observes a mixture of areas with coherent excitations (with finite spectral weight at the Fermi energy) and insulating regions (gapped spectra) as we would expect in our scenario. Upon heating the system becomes completely insulating (compare image at 320K). Note, after cooling down to 200K again the same "map" as before is recovered suggesting a correlation between the sites where the Cr–impurities sit and the insulating regions. This, however, is just a conjecture and has to be further clarified also theoretically.

The last part of our discussion about the optical conductivity is devoted to the data of the 1.1%Cr doped sample under pressure far in the metallic region at 6kPa. As it was stated in the beginning, and motivated by the results we already discussed, the second question we want to address is whether the doping with Cr can really be "reversed" by applying an external pressure. In short: Can the pressure be drawn on the same axis in the phase diagram as the doping? Experimental results from optical spectroscopy give a clear negative answer to that question. In Fig. 3.9 we report on the left hand side the comparison of the experimentally measured spectra for the undoped and the 1.1% Cr–doped sample. The spectra are, obviously, not even qualitatively similar. This indicates the existence of different PM states obtained by tuning temperature/doping or applying pressure.

It remains, however, to formulate and quantify this difference in a rigorous manner. This will be the subject of the next section, in which we discuss the hard x–ray absorption spectra on the vanadium K–edge.

3.1.4 X–ray absorption on the V–K–edge: Pressure vs. doping

The main part of the following discussion, despite the part about the linear dichroism, is published in the APS Journal "Physical Review Letters" [169]:
PRL **104**, 047401, (2010)

Among the different experimental methods recently employed to study the electronic properties of the Mott transition in Cr–doped V_2O_3 [114, 130, 131, 169, 168], X-ray absorption spectroscopy (XAS) has played a crucial role. For instance, it was the detailed investigation of the V $L_{2,3}$ absorption edges [153] that demonstrated the necessity of abandoning the simple one band, $S = 1/2$,

3 Bulk 3d–transition metal compounds: Theory vs. Experiment

model to obtain a realistic description of the changes in the electronic structure at the phase transition. Further, Park *et al.* obtained valuable quantitative information about the vanadium ground state for different amounts of doping and temperatures [153] and formulated it as a linear combination of the $|e_g^\pi e_g^\pi\rangle$ and the $|a_{1g} e_g^\pi\rangle$ states which were mentioned earlier. This kind of tool would be perfect to also clarify the question which remains from the discussion of the previous section: What is the character of the metallic ground state of the Cr–doped sample under pressure? However, unfortunately the V $L_{2,3}$ absorption falls in the region of soft x–ray radiation, and thus, due to the specific absorption characteristic of the diamond anvil cell used for the pressure measurements, it cannot be employed in our case. But fortunately XAS can also be performed at the V K–edge in the hard x–ray range, i.e., in a spectral region without particular absorption of the diamond anvil cell. In this case, the pre–edge will carry most of the physical information we are interested in, as it is predominantly due to $1s \rightarrow 3d$ transitions. The excitations in this pre–edge region are influenced by the core hole and should be considered to be of an excitonic nature. Beside the possibility of measuring the V K-edge under pressure condition we obtain also a more straightforward interpretation. Namely, due to the simple spherical symmetry of the s-core hole, the multiplet structure reveals a more direct view on the d-states.

Motivated by the above considerations, we used V K-edge XAS to explore extensively the MIT in V_2O_3 by changing temperature, doping and applying an external pressure. The onsets of the K-edges were analyzed by a novel computational scheme combining the LDA+DMFT method with configuration interaction (CI) full multiplet ligand field calculations (see last section in chapter 2) to interpret subtle differences at the PM–PI transition. This allowed us to: (i) observe in detail the changes in the electronic excitations, providing also a direct estimate of the Hund's coupling J (recall the discussion of the LDA+DMFT spectral functions in the previous chapter) (ii) analyze the physical properties of the PI and PM phase on both sides of the MIT, leading to the main result of our work: (iii) understand the difference between P, T or doping-induced transitions. This difference is mainly related to the occupancy of the a_{1g} orbitals, suggesting the existence of a new "pressure" pathway between PI and PM in the phase diagram. The XAS is in that

3.1 Optics and X–ray absorption of V_2O_3

respect complementary to the optical conductivity measurements. In chapter 2 we saw that the optical conductivity is connected to somewhat non–local excitations (therefore it was a great tool to confirm the mixed phase scenario). In contrast the XAS, or more specifically the excitonic features of XAS offers us information about the ground state from a completely localized perspective which is needed in order to formulate the ground state in the language of localized Wannier orbitals – this in turn is information which could not be extracted from the optical conductivity.

For the experiments we used high quality samples of $(V_{1-x}Cr_x)_2O_3$ with various doping in the PM ($x = 0$) and PI phases ($x = 0.011$ and 0.028) at ambient conditions. The MIT was also crossed for the 0.011 doping by changing temperature and for the 0.028 doping by pressure. To obtain the best resolution, the XAS spectra were acquired in the so-called partial fluorescence yield (PFY) mode [47], monitoring the intensity of the V-Kα ($2p \to 1s$) line as the incident energy is swept across the absorption edge. Further experimental details can be found in the published article [169]

powder data and isotropic calculations The T–dependent absorption spectra are displayed in Fig. 3.10 left hand side for both PM (200 K) and PI (300 K) phases for the $x = 0.011$ powder sample. The spectra have been normalized to an edge jump of unity. We will focus on the pre-edge region, where information about the V d-states can be extracted as it is indicated in the plot. It can be decomposed into three spectral features (A,B,C) which all vary in intensity as the system is driven through the MIT whereas C is considerably broader then A and B. Notice that no feature is observed below peak A contrary to the early results of Ref. [17] but in agreement with the more recent data of Ref. [72]. Within a simplified atomic like picture, one could directly relate the intensity of features A,B and C to the unoccupied states: The V–t_{2g}^2 states are split into one a_{1g} and two e_g^π states under trigonal distortion of the V sites [178] as shown in Fig. 3.3. Starting from a V–t_{2g}^2, $S = 1$ configuration, one can either add an electron to the t_{2g} subshell yielding peaks A and B, or add an electron to the e_g^σ sub–shell which gives rise to the broader peak C. In this picture, Hund's rule exchange splits peaks A and B into a quartet ($S = 3/2$)

3 Bulk 3d–transition metal compounds: Theory vs. Experiment

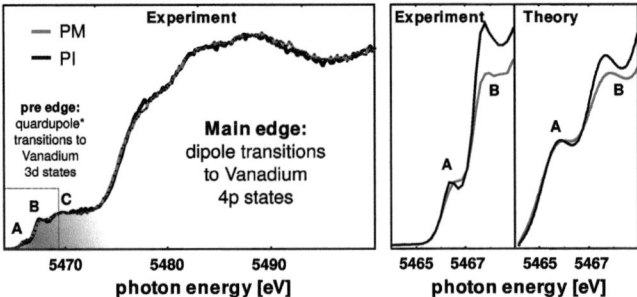

Fig. 3.10: Vanadium K-edge x-ray absorption spectra in $(V_{1-x}Cr_x)_2O_3$ for a powder sample with $x = 0.011$ measured as a function of temperature (T) in the PM (200 K, light gray line) and PI (300 K, dark gray line) phases by partial fluorescence yield XAS. In the region above 5475eV the main edge starts where dipole transitions from the core electron to the vanadium 4p states give the main contribution. Below 5470eV we find the pre–edge – here, besides others, the transitions to the vanadium 3d states are located. From these we extract the information about the ground state of the system. On the right hand side we show a zoom of the pre–edge region and compare the structure to theoretical full multiplet CI spectra. As explained in the text, these transitions would be pure quadrupole if it would not be for the inversion symmetry breaking on the vanadium site that makes the transitions "slightly dipole allowed".

3.1 Optics and X-ray absorption of V_2O_3

and doublet ($S = 1/2$) states.

This point of view is, however, oversimplified as the V d electrons are *strongly correlated* and, in the pre–edge region, the spectra are still largely influenced by the $1s$ core hole potential. Keeping that in mind, we have simulated the pre–edge by combining CI with LDA+DMFT calculations for which the one particle part (LDA) input corresponds to the level diagram in Fig. 3.3. We concentrate our analysis to peaks A and B, since peak C relates mainly to the unoccupied e_g^σ orbitals. These hybridize much stronger with the ligands and thus lack direct information on the Mott transition; peak C may also be related to non–local excitations (not included here) [71] which sensitively depend on the metal–ligand distance. Let us also note that the V sites in V_2O_3 are non centro–symmetric which leads to an on–site mixing of V-$3d$ and V-$4p$-orbitals and interference between dipole and quadrupole transitions [53]. This interference has been included in our scheme and will be discussed later in details in this section together with the linear dichroism measurements.

Our CI calculations confirm that for the ground state the occupancy ratio between the (e_g^π, a_{1g}) and (e_g^π, e_g^π) states is smaller in the PI than in the PM phase [102, 153]: The isotropic CI–based calculated XAS spectra in the pre-edge region reported in Fig. 3.10 right hand side agree well with the experimental data for both the energy splitting of features A and B and the ratio of their spectral weight (SW) which increases in the PM phase.

Considerable insight can be gained by comparing multiplet CI and LDA+DMFT calculations. Our LDA+DMFT calculations, performed using the same NMTO Hamiltonian as discussed in section 3.1.2 with the 1.1% Cr–doped V_2O_3 and Hirsch–Fye Quantum Monte Carlo as impurity solver, confirm the above mentioned tendency. Specifically we obtain a mixing of 50:50 and 35:65 for the $(e_g^\pi, a_{1g}):(e_g^\pi, e_g^\pi)$ occupation in the PM and PI phases respectively. Remarkably the simple structure of the core hole potential in the K-edge spectroscopy ($L = 0$ i.e. spherical potential) allows us to associate the pre–edge spectrum with the k–integrated spectral function above the Fermi energy calculated by LDA+DMFT in which we do not take into account the core hole effects. The electron–addition part of the spectral function shows three main features in PM phase: a coherent excitation at the Fermi level and a much broader dou-

3 Bulk 3d–transition metal compounds: Theory vs. Experiment

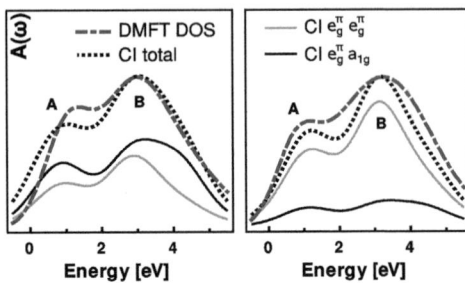

Fig. 3.11: Incoherent LDA+DMFT and CI calculations in the pre-edge region; ε_F is the Fermi energy. Note the similarity in the main spectral features when crossing the MIT. Also shown are the different contributions of the CI spectrum labeled accordingly to their initial state: the contribution of the $(e_g^\pi, a_{1g}) \to (e_g^\pi, e_g^\pi, a_{1g})$ transitions to the peak A(B) is approximately 60%(55%) in the PM phase and 20%(15%) in the PI phase.

ble peak associated to the incoherent electronic excitations, i.e., the upper Hubbard band (UHB), almost identically to the undoped compound. In the PI phase obviously, only the latter survives. Comparison with the experimental spectra clearly shows that the pre–edge features have to be related to the "incoherent" part of the spectral function only. The physical reason is that the core hole potential localizes the electrons destroying the (already strongly renormalized) coherent quasiparticle excitations and making the XAS spectrum atomic–like (see also discussion on XAS at the end of chapter 2). All the "incoherent" LDA+DMFT, CI, and experimental spectra shown in Fig. 3.11 agree in many aspects, especially as for the splitting of the first two peaks by ≈ 2.0 eV (≈ 1.8 eV in experiment) which originates in LDA+DMFT from the Hund's exchange J in the Kanamori Hamiltonian (see discussion of the LDA+DMFT spectral functions on page 60). This further validates the choice of $J = 0.7$ eV used in our calculations in contrast to larger values assumed in previous studies [102, 110], and also clarifies the mismatch between XAS and LDA+DMFT spectra reported in the undoped V_2O_3 compound [102] where incoherent excitonic features were identified by coherent quasiparticle excitations. Moreover, the ratio between A and B peak displays the same trend

3.1 Optics and X-ray absorption of V_2O_3

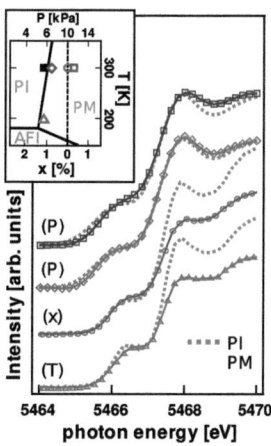

Fig. 3.12: V–K edge XAS spectra for powder samples of $(V_{1-x}Cr_x)_2O_3$, starting from the top, as a function of pressure (P) (□) [$x = 0.028$; 5 and 11 kbar (topmost set of curves); 5 and 7 kbar (second highest set of curves)], temperature (△) [$x = 0.011$; 200,300 K (second lowest set of curves)] (T), and doping (∘) [$x = 0$, 0.011 (bottom curves)] (x) (cf. points in the phase diagram; the pressure scale refers to the $x = 0.028$ doping). The spectral differences demonstrate the nonequivalence between P and temperature-doping. The x-T equivalence is confirmed by the photoemission spectra [170].

in the PM-PI transition as the CI (or experimental) data. The quantitative difference between the two calculations is attributed to the lack of matrix elements in LDA+DMFT.

The intensity ratio of the first two incoherent excitations peaks A and B (associated to the quartet and doublet states in the oversimplified picture) thus appears as the key spectral parameter to understand the differences between PM and PI. Even in a powder sample, this ratio is still sensitive to the a_{1g} orbital occupation of the initial state (in the last paragraphs of this section we will see that polarization dependent measurements give an even clearer picture). Indeed, due to the trigonal distortion a considerable spectral weight transfer from the peak B to higher energies (corresponding to final states with two a_{1g} electrons in the limit of large Δ_{trig}) can take place for the (e_g^π, a_{1g}) but not for the (e_g^π, e_g^π) initial state. Therefore, the K pre–edge XAS can serve as a direct probe of the a_{1g} orbital occupation in the ground-state. As a rule of thumb, the larger the ratio between the SW of A and B, the larger the a_{1g} orbital occupation.

under pressure Now, after we established an interpretation scheme of the vanadium XAS K–edge which allows us to use it as a ground state probe it is

3 Bulk 3d–transition metal compounds: Theory vs. Experiment

time to come back to the original task of inquiring the metallic phase of Cr–doped V_2O_3. Fig. 3.12 shows the XAS powder spectra of the pressure–induced MIT with the corresponding spectra for the temperature– and doping–driven transition (the markers in the phase diagram, Fig. 3.12). We remark at this point that the spectra taken under pressure display a relative shift between main–edge (not shown) and pre–edge, which is in any case irrelevant for our discussion of the ground state for which we only need the intrinsic structure of the pre–edge. Hence, this shift is compensated for the pressure spectra in Fig. 3.12. Fig. 3.12 clearly evidences that (besides the shift) contrary to the doping- or T-driven transition, very small changes in spectral shapes and weights are observed in the pressure driven MIT. In the light of the arguments discussed above, our finding proves that the metallic state reached by applying pressure is characterized by a much lower occupation of the a_{1g} orbitals compared to the metallic state reached just by changing temperature or doping. Importantly, the spectra measured through the doping induced MIT are identical within the experimental uncertainty to those measured through the temperature driven transition. The temperature–doping equivalence is confirmed by photoemission data [170] and is consistent with the very similar lattice parameter changes across the transition [125]. The x and T equivalence is also borne out by the observation from XAS at the L-edges in doped V_2O_3 [153] that the a_{1g} occupation within both the PM or PI phases is mostly independent of the doping level. Hence, the local incoherent excitations probed by XAS at the V L edge or K pre-edge are not directly affected by disorder [65]. The reason for this is that XAS is a *local probe* in the sense that we can expect the changes in the XAS spectrum to be of the order of the percentage of the atoms which have a different ground state: upon disturbing 1% of the atoms due to Cr doping we expect changes in the spectrum of the order of 1% (while a perturbation of 1% can lead to the total breakdown of a picture in momentum space).

Our finding clearly shows the limits of the common assumption that temperature, doping, and pressure–driven MITs in V_2O_3 can be equivalently described within the same phase diagram [125][7]. Indeed, the two different PM elec-

---

[7] An early version of the phase diagram (Fig. 15 in [125]) was actually drawn with a third pressure axis, but due to the idea of p–x equivalence, this was later abandoned.

3.1 Optics and X-ray absorption of V_2O_3

tronic structures that we observed reflect different mechanisms driving the MIT along different pathways. In the doping–driven MIT, the metallic phase is characterized by an increased occupation of the a_{1g} electrons indicating a reduced "effective crystal-field-splitting" as the main driving mechanism towards metallicity [102, 157], related to the jump of the lattice parameter c/a (1.4%) at the MIT [125]. In contrast, when pressure is applied, the a_{1g} occupation remains basically unchanged, so that this metallic phase seems to originate rather from an increased bandwidth, without any relevant changes of the orbital splitting. The smaller c/a jump observed under pressure (0.7 %) corroborates our analysis.

In conclusion, doping, temperature and pressure are shown to act differently on the interplay between electron correlations and crystal field, so that states previously considered to be equivalent metals are actually different.

linear dichroism: an introduction In the previous discussion we concluded, that the ratio between the first two peaks A and B of the V K–pre–edge can be utilized as a robust ground state probe. As we have mentioned before, the sensitivity of this probe can be strongly enhanced by using polarized x–ray light with single crystals instead of powder samples. In this case the absorption spectrum depends also on the specific polarization of the x–ray light and, consequently, carries more information. Polarization dependent absorption is a well known effect and can even be observed in everyday life, e.g., in the glasses for 3D cinema or polfilters for photography. In general this effect is called dichroism and can be further subcategorized in circular (CD) and linear (LD) dichroism, depending on the polarization of the probe light. In the last fifteen years dichroism measurements in XAS have proven to be a powerful tool:

Circular dichroism of spectra at the L–edge of transition metals were measured for the first time in the nineties[35, 36, 199, 200] and calculated afterwards by van der Laan and Thole [208]. Exploiting sum–rules, which connect the spin– and orbital–magnetic moments with the integrated inten-

3 Bulk 3d–transition metal compounds: Theory vs. Experiment

sity [197, 26], the CD XAS has evolved to a widely used tool for "spectroscopic susceptibility measurements".

Linear dichroism, on the other hand, is caused by non–cubic orbital occupation and has been utilized with great success as crystal field ground state probe [153, 37, 141, 77, 73, 74]. Hence, it presents a possible additional resource of information for our V_2O_3 analysis. The reason for linear dichroism and its relation to the occupancy of crystal field eigenstates can be understood intuitively even in a one–electron picture. Consider the following example: We want to promote a core s–electron via a dipole transition to a valence state of p_z–symmetry. Following Fermi's golden rule in the sudden approximation, the spectrum of such a process is proportional to the square of the integral $\langle s|\hat{D}|p_z\rangle$ with $\langle s| \propto 1$ and $|p_z\rangle \propto z$ in Cartesian coordinates. If the integrand is an odd function the integral vanishes. Hence, the *only* dipole transition operator for which the integral remains finite is the $\hat{D}_z \propto z$. The same holds, of course, also for the corresponding transitions to p_x– or p_y–states, i.e., with x and y polarized light one can excite into the p_x– or p_y–states respectively. Now, let the ground state of our hypothetical system be a single occupied p_z–orbital and empty p_x and p_y orbitals. Hence, adding an electron to the p_z–orbital will cost additional coulombic energy, so that the peak of the z–polarized spectrum (corresponding to the $\hat{D}_z$ transition operator) lies at higher energies compared to the $\hat{D}_{x/y}$ spectra showing also only half the intensity due to its occupancy. Consequently, we find valuable information about the ground state of the system encoded in the LD spectra. In other words: The polarization of the x–ray light adds *more selection rules* to the absorption process which we can use for a more detailed study of the ground state properties.

Let us recall the most important transitions for the V K–edge: As also indicated in Fig. 3.10 on the left hand side, the main edge consists mostly of dipole transitions of the s–core electron into the unbound V 4p–states, whereas transitions to the localized and correlated 3d–states via higher order processes contribute to the pre–edge. We formally write the absorption

3.1 Optics and X–ray absorption of V_2O_3

process employing Fermi's golden rule as:

$$\left| \sum_{\Psi_{\text{final}}} \langle \Psi_{\text{final}} | \hat{P} | \Psi_{\text{initial}} \rangle \right|^2 \delta(\omega - E_{\text{initial}} + E_{\text{final}}) \tag{3.5}$$

This quantity is directly related to the spectral intensity. The transition operator $\hat{P}$ describes the coupling of the probe light to our system and can generally be written as [115, 127]:

$$\hat{P} = e^{i\mathbf{k}\cdot\mathbf{r}} \mathbf{p} \cdot \hat{\mathbf{e}} \tag{3.6}$$

where **k** and **ê** are the light propagation and polarization vectors and **r** and **p** are the electron position and momentum operators.

We expand (3.6) around the limit of long wavelengths (i.e. $2\pi/\lambda \ll 1$) and obtain (also using $\mathbf{p} = im/\hbar[\hat{H}, \mathbf{r}]$) as first element the dipole operator $\hat{D} = \hat{\mathbf{e}} \cdot \mathbf{r}$, which can be expressed in terms of spherical harmonics Y_k^m:

$$\begin{aligned}
\hat{D}_x &= \frac{r}{\sqrt{2}}(Y_1^{-1} + Y_1^1) \\
\hat{D}_y &= \frac{ir}{\sqrt{2}}(Y_1^{-1} - Y_1^1) \\
\hat{D}_z &= rY_1^0
\end{aligned} \tag{3.7}$$

The second element in the expansion of (3.6) is the quadrupole transition operator $\hat{Q} = (\hat{\mathbf{e}} \cdot \mathbf{r})(\mathbf{k} \cdot \mathbf{r})$, or in terms of the Y_k^m

$$\begin{aligned}
\hat{Q}_{xy} &= \frac{ir^2}{\sqrt{2}}(Y_2^2 - Y_2^{-2}) & \hat{Q}_{x^2-y^2} &= \frac{r^2}{\sqrt{2}}(Y_2^2 + Y_2^{-2}) \\
\hat{Q}_{xz} &= \frac{r^2}{\sqrt{2}}(Y_2^1 - Y_2^{-1}) & \hat{Q}_{3z^2-r^2} &= r^2 Y_2^0 \\
\hat{Q}_{yz} &= \frac{ir^2}{\sqrt{2}}(Y_2^{-1} + Y_2^1)
\end{aligned} \tag{3.8}$$

3 Bulk 3d–transition metal compounds: Theory vs. Experiment

The transitions (3.5) and, thus, also the absorption cross–section are proportional to Clebsch–Gordan like integrals as they were shortly discussed in the small s–core to p_z–state example above. They will provide all the LD selection rules encoding the ground state information we seek. Let us now turn to the linear dichroism of the vanadium K–edge.

linear dichroism in the vanadium K–edge In Fig. 3.13 we show the experimental spectra (dots) of the V K–edge for three different polarizations of the x–ray light. On the right hand side, we sketch the polarization of each spectrum in terms of the trigonal (local vanadium) reference frame (a, b, c) [8]. Since we want to formulate the transition operators for each polarization in terms of the dipole and quadrupole operators (3.7) & (3.8), we translate to Cartesian coordinates:

$$a = x \qquad\qquad x = a$$
$$b = \frac{\sqrt{3}}{2}y - \frac{1}{2}x \qquad\qquad y = \frac{1}{\sqrt{3}}(a + 2b)$$
$$c = z \qquad\qquad z = c$$

Now we can formulate the dipole and quadrupole operators for each orientation in terms of (3.7) & (3.8)

$$\hat{D}_b = \hat{D}_x \qquad\qquad \hat{Q}_b = \hat{Q}_{yz}$$
$$\hat{D}_g = \frac{\sqrt{3}}{2}\hat{D}_y + \frac{1}{2}\hat{D}_x = \hat{D}_b \qquad\qquad \hat{Q}_g \equiv \hat{Q}_{x^2-y^2}$$
$$\hat{D}_r = \hat{D}_z \qquad\qquad \hat{Q}_r = \frac{\sqrt{3}}{2}\hat{Q}_{xz} + \frac{1}{2}\hat{Q}_{yz} \equiv \hat{Q}_b$$

---

[8] where $\hat{a}$ and $\hat{b}$ form an angle of 120 degree and are both perpendicular to $\hat{c}$ (compare Fig. 3.13)

3.1 Optics and X-ray absorption of V_2O_3

(Here we used the symmetry $a \leftrightarrow b$ of the crystal).
Evidently, there are symmetry relations between the transition operators for the three different polarizations. While for dipole transitions we find the lightgray and darkgray polarization to be equivalent, in quadrupole transitions lightgray and black polarization should be indistinguishable.
However, we find the symmetries obviously violated when we inspect the spectra closely: Inset (I) shows a zoom of the pre–edge region where clearly all three spectra are different. Moreover, even in the main edge we find the "lightgray/darkgray" dipole symmetry violated. The reason for this violation is, in fact, quite simple and can be understood intuitively with the sketch of the primitive unit cell (Fig. 3.2 right hand side). Along the connection line of two vanadium atoms in the primitive unit cell, i.e., parallel to the crystallographic c–axis the vanadium site lacks inversion symmetry, since the distances to the "upper" and "lower" V–neighbors is different[9]. The consequence of breaking such a fundamental symmetry is quite dramatic, namely the V 3d–states are no longer pure eigenstates of the system and start to mix with V 4p–states *onsite*. In other words this means, that the formerly pure d–states get a tiny dipole moment and the quadrupole/dipole selection rule no longer holds strictly. This mixing will explain why the pre–edge is not entirely quadrupole in character and the main–edge not entirely dipole.

Let us write down what consequences the onsite d–p mixing will have for the transitions (3.5) in a simplified fashion. The final state is now of a mixed V 3d and V 4p nature, which we express as a linear combination of a part with V 4p and one with V 3d character $\alpha|\Psi_{4p}\rangle + \beta|\Psi_{3d}\rangle$. Hence we write Eq. (3.5)

---

[9] A beautiful visualization of this situation can be found in the Wannier function plots in Fig. 15 of the work by Saha–Dasgupta *et al.* [178]

3 Bulk 3d–transition metal compounds: Theory vs. Experiment

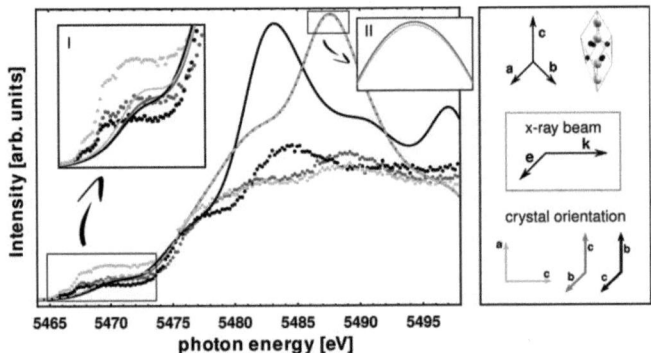

Fig. 3.13: Polarization dependent experimental data for the vanadium K–edge (points) compared to calculated LDA data. The color of the respective plot codes the geometry of the measurement, i.e., the transition operator, explained in the sketch on the right hand side. We can see the effects of the inversion symmetry breaking along the V–V bonds nicely in insets I and II, where we observe the splitting of spectra which should be equivalent by definition in a system without the symmetry breaking. The differences in the intensities of the largest peaks can be attributed to self absorption effects, which are not included in the calculation.

3.1 Optics and X-ray absorption of V_2O_3

as:

$$\sum_{\Psi_f} \left| \langle \Psi_i | \hat{P} | \Psi_f \rangle \right|^2 \delta(\omega - \Delta E) = \sum_{\Psi_f} \left| \langle \Psi_{1s}^{\text{core}} | (\hat{D} + \hat{Q})(\alpha | \Psi_{4p} \rangle + \beta | \Psi_{3d} \rangle) \right|^2 \delta(\omega - \Delta E)$$

$$= \sum_{\Psi_f} \left| \langle \Psi_{1s}^{\text{core}} | (\alpha \hat{D} | \Psi_{4p} \rangle + \beta \hat{Q} | \Psi_{3d} \rangle) \right|^2 \delta(\omega - \Delta E) \quad (3.9)$$

where we set $\Delta E = E_f - E_i$. Now the spectral intensity for the absorption is proportional to the square of (3.9). However, due to the mixed character of the final state $|\Psi_{4p}\rangle + |\Psi_{3d}\rangle$, we cannot write the total spectrum as a sum of a dipole spectrum and a quadrupole spectrum. Both parts are entangled by interference terms of the form

$$\alpha \beta \langle \Psi_{1s}^{\text{core}} | \hat{Q} | \Psi_{3d} \rangle \cdot \langle \Psi_{1s}^{\text{core}} | \hat{D} | \Psi_{4p} \rangle \quad (3.10)$$

which have to be taken into account explicitly.

For the main edge, i.e., where excitonic features and the many–body coupling to the core hole are negligible, this is feasible within the LDA. Since the *ab initio* LDA calculation incorporates the symmetry breaking in the structural input, the information about the resulting d–p mixing is already *included* in the eigenstates of the LDA Hamiltonian.

In order to calculate the XAS spectrum we have to consider only the muffin tin sphere around the vanadium atom R_{MT} since the V 1s core hole wave function is zero for $r > R_{\text{MT}}$. We calculate projections of the LDA wave functions on the V 3d and V 4p subspace within the sphere:

$$\Psi_{l,m}^{\text{LDA}} = \mathcal{D}_{l,m}(\omega) R_{lm}(r) Y_l^m \quad (3.11)$$

where we neglected the energy dependence of the radial part of the wave functions, and the energy dependent $\mathcal{D}_{l,m}(\omega)$ stems from the projection. The square of $\mathcal{D}_{l,m}(\omega)$ is in fact the so called *partial density of states*.

In terms of (3.11) the integrals of (3.9) separate into three parts: i) partial DOS $\mathcal{D}^2_{3d(4p)}(\omega)$ and "interference terms" $\mathcal{D}_{3d}(\omega) \cdot \mathcal{D}_{4p}(\omega)$ ii) Radial transition probabilities of the form

$$\int_0^{R_{\mathrm{MT}}} r^2 dr \cdot r R^*_{1s}(r) R_{4p}(r) \tag{3.12}$$

for the dipole part, and

$$\int_0^{R_{\mathrm{MT}}} r^2 dr \cdot r^2 R^*_{1s}(r) R_{3d}(r) \tag{3.13}$$

for the quadrupole part. And last iii) angular integrals like

$$\int d\Omega\, Y_l^m O_{l'}^{m'} Y_{l''}^{m''} \tag{3.14}$$

where $O_{l'}^{m'}$ is the angular part of the specific transition operator. The angular integrals yield the selection rules for the transition.

The solid (lightgray dashed) lines in Fig. 3.13 represent the LDA spectra for the respective (color coded) transition symmetry. The overall agreement for the main edge is very satisfying. Especially the interference terms between 3d and 4p states, which are captured by the calculation yield interesting features. As a signature of this mixing we find a splitting of the three spectra (see insets I and II) in the experimental and calculated data which would not be present in pure dipole/quadrupole transitions.

In the pre–edge region the LDA calculation misses, of course, the excitonic features (the first two peaks) which simply involve many–body states beyond LDA. Nonetheless, the LDA captures well the correct trends in the pre–edge region around 5470eV where transitions to the e_g^σ, the least correlated and most itinerant among d–states, are located. In order to utilize the ratio of the first two excitonic peaks to probe the $\alpha|e_g^\pi, e_g^\pi\rangle + \beta|e_g^\pi, a_{1g}\rangle$ ($\alpha^2 + \beta^2 = 1$) ground state, however, we have to include the 3d–4p mixing in our cluster calculation. This can be done by taking this mixing, i.e. the overlap integrals of V 3d

3.1 Optics and X-ray absorption of V_2O_3

and V4p, as a fitting parameter[10] and calculate also the interference terms (3.10). As a technicality we remark that the t_{2g} and e_g^σ spectra were calculated separately and summed up later. This was done in order to account for the big differences we expect for the broadening of the excitations due to the much stronger non–local hybridization with the e_g^σ ligands[11]. Moreover, we calculated the spectra for all four vanadium atoms in the unit cell and averaged them[12].

In Fig. 3.14 we report the results of our CI together with experimental data for the PM and PI phase. On the left hand side of the figure we show a map of simulated spectra for different values of α^2. Overall we observe that we can capture the trend of the spectrum hierarchy also in the cluster calculation quite well. This time, however, the itinerant states are beyond the basis set of our CI calculation of a small cluster and we cannot expect quantitative agreement for energies in the region of the e_g^σ. On the other hand, we now capture the excitonic peaks A and B correctly which we need as a ground state probe. Let us remark at this point, that the e_g^σ–states can be captured from both sides, LDA and CI, only qualitatively. In fact, they can be understood to be placed in a kind of intermediate state – neither truly localized nor totally itinerant.

In the middle panel and in the panel on the right hand side of Fig. 3.14 we show the experimental spectra compared to the theoretical CI results which we obtained with the same mixing parameters we used for the isotropic spectra, i.e., very close to the DMFT results. The agreement between experiment and CI calculations is quite satisfying. Moreover, we observe that the change of the ratio between peak A and B, i.e., our most important probe, going from PM to PI is much more pronounced in the selected LD spectra compared to the powder spectra: In the isotropic case the spectral weight transfer only took place for peak B to higher energies, yet, in the LD data additional se-

---

[10] The proper *symmetry* of the potential we take from a Madelung calculation
[11] Further, as it was stated also before, the mixing of the t_{2g} and the e_g^σ are neglected.
[12] Locally the vanadium atoms are equivalent, but not with respect to an external frame of reference like the propagation and polarization vector of the x-ray light.

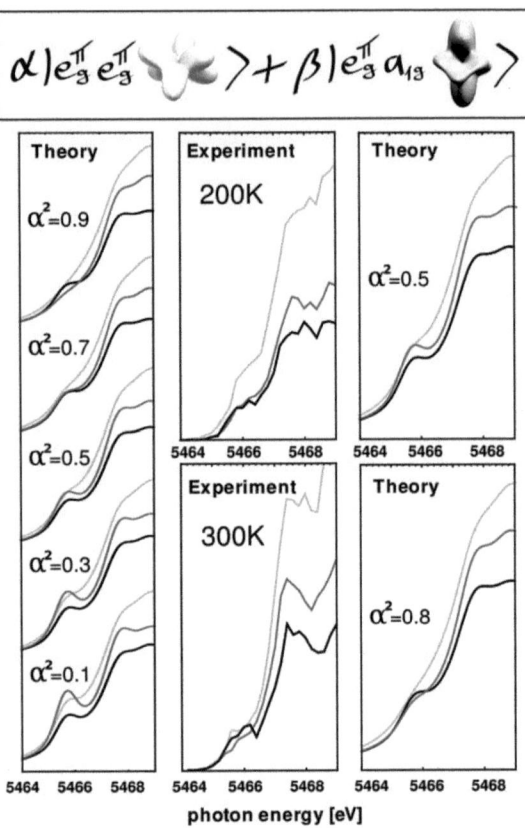

Fig. 3.14: Linear dichroism in the pre–edge region of the vanadium K–edge. In the panel on the left hand side we show a "map" of theoretically calculated full multiplet spectra for different compositions of the ground state $\alpha|e_g^\pi\rangle + \beta|a_{1g}\rangle$ (with $\alpha^2 + \beta^2 = 1$). Employing this map we find the best agreement with the experimental data (middle panel) for values of $\alpha^2 = 0.5$ for the PM phase (200K) and $\alpha^2 = 0.8$ for the PI phase (300K). Hence, also here we find that the LDA+DMFT densities are a good starting point for the CI calculation.

lection rules affect peak A directly, which has spectral weight *almost* proportional to the amount of $|e_g^\pi, a_{1g}\rangle$ character in the ground state. This remarkable effect is, of course, due to LD selection rules and in a way just the same as the toy–example of the p_z ground state discussed at the beginning of the LD paragraph. It can be understood easiest in the limit of a pure quadrupole transition, where we neglect the mixing of the V 3d–states with the V 4p–states. In this limit the first peak is for our transition operators, in fact, displayed *exclusively* in the $|e_g^\pi, a_{1g}\rangle$ part of the spectrum and completely absent in the $|e_g^\pi, e_g^\pi\rangle$ part. The transitions associated with this peak is the $|1s_{\text{core}}\rangle \to |e_g^\pi, e_g^\pi, a_{1g}\rangle$. If the ground state is $|e_g^\pi, a_{1g}\rangle$ the 1s–core electron has to be placed in the free e_g^π level which can be done. If, however, the ground state is $|e_g^\pi, e_g^\pi\rangle$ the 1s–core electron has to be placed in the free a_{1g} level – but this is impossible with the specific transition operators of the displayed spectra, since all the associated matrix elements are integrals over odd functions and thus zero. Hence, we are sensitive to the a_{1g} occupation of the ground state. Let us stress that this argument holds strictly only in the abovementioned assumption of no 3d–4p mixing. In the real situation, and also our full calculation, this mixing weakens this effect slightly (the first peak is displayed also in the $|e_g^\pi, e_g^\pi\rangle$ part of the spectrum due to the d–p–hybridization) but nonetheless it is still observable. In conclusion, the experimental LD data of Fig. 3.14 presents further evidence for the robustness of our ground state probe and their theoretical simulation is consistent with the ground state parameters we found for the isotropic case.

3.2 Optics of NiSe$_x$S$_{1-x}$

The main part of the following discussion is published in the APS Journal "Physical Review B" [155]: PRB **80**, 073101 (2009)

In the first section of this chapter we discussed one of the prototypes of correlated systems: Vanadium sesquioxide V$_2$O$_3$. The transition metal oxides

3 Bulk 3d–transition metal compounds: Theory vs. Experiment

present, as stated before, one of the largest families in the area of correlated solid state physics. Moreover, oxygen has a high electronegativity which causes its transition metal compounds often to be of a rather ionic nature. This ionicity, in turn, leads to a separation of oxygen and transition metal degrees of freedom so that the 3d–bands around the Fermi energy are hybridized only very little with the oxygen 2p–bands. This situation is a good starting point for downfolding/projecting to a low energy model. In the following discussion we leave the ground of the oxides and take a step down in the periodic table to sulfur and selenium, where we find a different situation. The experimental data we show in this section was obtained by the group of Prof. S. Lupi in the University "La Sapienza" in Rome.

The cubic Vaesite NiS_2 (pyrite structure), which is a charge transfer (CT) insulator following the Zaanen-Sawatsky-Allen classification scheme [220], is also considered a text–book example of a strongly correlated material. NiS_2 attracts particular interest since it easily forms a solid solution with $NiSe_2$ ($NiS_{2-x}Se_x$), which, while being iso–electronic and iso–structural to NiS_2, is nevertheless a good metal. A metal to insulator transition, induced by Se alloying, is observed at room temperature for $x \approx 0.6$. Further, a magnetic phase boundary between an antiferromagnetic–metal to a paramagnetic–metal is found at low temperature at about $x = 1$ (see the inset of Fig.3.16a) [89].

An alternative way to induce a metallic state in NiS_2 is by application of a hydrostatic pressure. Following Mott's original idea [135], this procedure, although often technically challenging, the unique opportunity to continuously and homogeneously tune the band–width W, without introducing impurities or disorder. High pressure techniques have indeed been used in the past few years to investigate the dc transport properties of $NiS_{2-x}Se_x$ [129, 140], and a pressure induced MIT has been observed in pure NiS_2 for $P > 4$ GPa.

As we have seen in the previous section, optical reflectivity/conductivity is a very suitable tool for probing the electronic structure of strongly correlated systems. However, with few relevant exceptions [150, 41], infrared investigations of the metal to insulator transition in strongly correlated charge transfer insulators are still rare. In this section we discuss room temperature optical

3.2 Optics of NiSe$_x$S$_{1-x}$

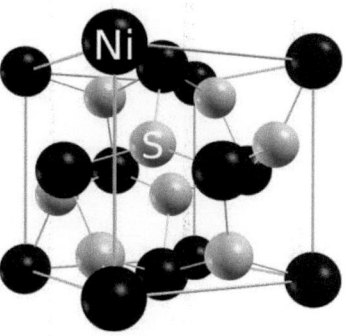

Fig. 3.15: Conventional unit cell for the crystal structure of nickel disulfide NiS$_2$. In the so called pyrite structure nickel is coordinated by undistorted octahedra of Sulfur ligands. The two Sulfur atoms in the center of the conventional unit cell form strong $\sigma-$ bonding/antibonding states, of which the antibonding are important degrees of freedom at the Fermi energy.

reflectivity measurements over a broad spectral range on 4 compounds ($x = 0$, 0.55, 0.6, 1.2) of the NiS$_{2-x}$Se$_x$ series together with optical measurements as a function of pressure on pure NiS$_2$. The experimental data, compared with LDA calculations, shows that the two MITs observed as a function of pressure and Se alloying actually rely on different microscopic mechanisms.
For details on the measurement we refer to the published article as well as [13, 10, 155].

experimental data: MIT for Se doping (x) & pressure (P) Let us start with the presentation of the experimental data. In the upper panel of Fig.3.16 we show the ambient pressure reflectivity $R(\omega)$ of NiS$_2$. We observe a nearly flat behavior from 50 to 10000 cm^{-1} except for small phonon modes at about 260 and 290 cm^{-1}. On increasing the Se–content, the reflectance is progressively enhanced at low frequencies, with a $R(\omega) \to 1$ for $\omega \to 0$ which is characteristic of the metallic behavior. Let us, however, perform a Kramers–Kronig transformation[13] and analyze the MIT by means of the optical conductivity. In Fig. 3.16 we plot the real part of the optical conductivity $\sigma_1(\omega)$. For NiS$_2$ ($x = 0$, black curve) we observe a clear insulating shape of $\sigma_1(\omega)^{x=0}$: it is strongly depleted at low frequency, showing the (CT) gap of about 4000 cm^{-1} which is consistent with previous optical measurements [101, 129]. Upon

---

[13]for the KK transformation we adopted standard extrapolation procedures both at high and low frequency [217, 51].

3 Bulk 3d–transition metal compounds: Theory vs. Experiment

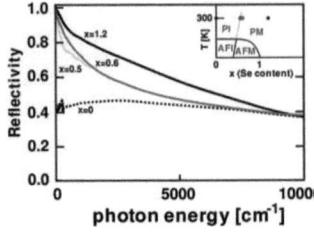
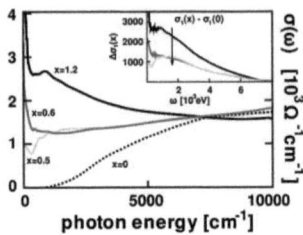

Fig. 3.16: Left hand side: Experimentally measured optical reflectivity of $NiS_{2-x}Se_x$ for $x = 0$, 0.55, 0.6, 1.2 at ambient conditions. Inset: Phase diagram of $NiS_{2-x}Se_x$ [89] with black dots corresponding to the samples measured in the present work. Right hand side: Optical conductivities from KK transformations. Inset: Difference $\Delta\sigma_1 = \sigma_1(x) - \sigma_1(x = 0)$ spectra.

increasing the doping with Selenium ($x > 0$) we observe the appearance of a large amount of spectral weight at low frequency. Further, we find an indication that electronic correlations are playing an important role in the MIT of $NiS_{2-x}Se_x$ observing an isosbestic point around 8000 cm^{-1}, through which the SW is transferred from higher to lower frequency. Such a large energy scale arises from the large–energy quantities that govern the opening and closing of the CT gap, namely, 3d electron correlation and the p–d transfer interaction.

As it is better highlighted by the $\Delta\sigma_1 = \sigma_1(x) - \sigma_1(x = 0)$ difference spectra, shown in the inset of the lower panel of Fig.3.16, the low energy contribution is made up of two well distinct terms: one broad mid–infra red feature peaked around 2000 cm^{-1} and extending up to nearly ≈ 8000 cm^{-1} and a sharp term below 500 cm^{-1}. The narrow peak can be attributed to coherent transport of quasi–particles around the Fermi energy, while the mid–infra red term is associated to optical transitions between the quasi–particle peak at ε_F to the upper and lower Hubbard bands.

We now turn to the high–P measurement of NiS_2. We show the reflectivity at the sample–diamond interface $R_{sd}(\omega)$ in Fig.3.17. The strong two–phonon diamond absorption allows to obtain reliable $R_{sd}(\omega)$ only above 2000 cm^{-1}. On increasing the pressure $R_{sd}(\omega)$ is progressively enhanced at low frequency

3.2 Optics of $NiSe_xS_{1-x}$

showing an overdamped behavior, as a signature for a correlated bad metallic state. At high frequencies, all $R_{sd}(\omega)$ converge above 10000 cm^{-1}. In order to evaluate the accuracy of our high–P measurements, we calculate the expected reflectivity at a sample–diamond interface, $R_{sd}^{cal}(\omega)$, by using a procedure previously introduced in [13, 10]. The calculated $R_{sd}^{cal}(\omega, x = 0)$ for NiS$_2$ (Fig.3.17 is in good agreement with $R_{sd}(\omega)$ measured in the diamond anvil cell at the lowest pressure (1.1 GPa) being both nearly flat and with a value $\approx 20\%$ over the whole frequency range. The same calculation has been carried out for NiS$_{2-x}$Se$_x$ ($x = 0.55$, 0.6, and 1.2) compounds and the resulting $R_{sd}^{cal}(\omega)$ are shown in Fig.3.17 for comparison. Evidently that the $R_{sd}(\omega)$ of NiS$_2$ measured on increasing P resemble the $R_{sd}^{cal}(\omega)$ obtained on varying x.

The microscopic mechanisms inducing the P and Se MITs are further investigated by studying the quasiparticle spectral weight as a function of the cubic lattice parameter a. The lattice is expanded by Se alloying [66, 109] whereas it is compressed by pressure [196]. The x and the P dependence (up to ~ 5 GPa) of a have been obtained from Refs. [109] and [196], respectively. Data at higher P have been extrapolated [177]: Through the specific heat results of Ref.[218] we obtain a sound velocity $v_s \approx 4300$ m/s. Since the density of NiS$_2$ is $\rho = 4455$ kg/m^3, the Bulk modulus results $B_0 = \rho \cdot v_s^2 \approx 83$ GPa. One usually assumes a linear dependence of the bulk modulus $B(P) = B_0 + B'P$, and $a(P)$ is finally given by the Birch–Murnagham (BM) equation [139]

$$a(P) = a(0) * \left[1 + \frac{B'}{B_0} * P\right]^{-1/3B'} \quad (3.15)$$

where B' normally ranges between 4 and 8 [97]. Measured $a(P)$ data together with values obtained from equation (3.15) are presented in the inset of Fig. 3.17. We remark that experimental data from Ref. [196] well agree with those obtained from equation (3.15). The overall decreasing behavior of a is consistent with that suggested by LDA lattice parameter calculations (see below) also shown in the same inset. The pressure vs. lattice constant was calculated using the TB–LMTO–LDA method and the co–called *force theorem*[14],

---
[14] Calculations performed by Ove Jepsen (MPI–Stuttgart)

3 Bulk 3d–transition metal compounds: Theory vs. Experiment

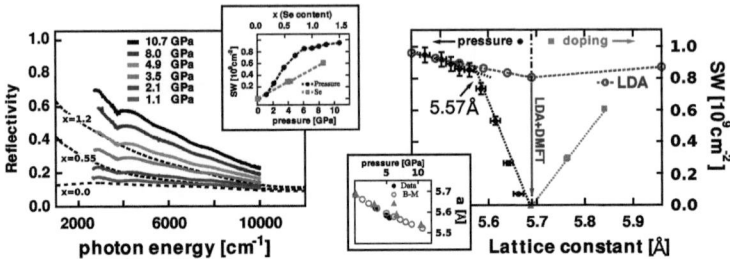

Fig. 3.17: Left hand side: Experimentally measured high pressure $R_{sd}(\omega)$ (thick solid lines) for NiS$_2$ and calculated $R_{sd}^{cal}(\omega)$ (dashed lines) at sample–diamond interface for NiS$_{2-x}$Se$_x$ at selected x. Right hand side: the QP low frequency spectral weight versus the lattice parameter a for pure NiS$_2$ under pressure and for NiS$_{2-x}$Se$_x$. Dashed lines are a guide to the eye. The dashed–dotted vertical line marks the lattice parameter value for NiS$_2$ at ambient conditions. Inset: Lattice parameter a as a function of P: experimental data from Ref.[196] (solid circles), calculated values using the B-M equation (open circles) and LDA (solid triangles).

which is based on the change of the total energy with uniform compression, see e.g. [116]. We show the low frequency quasiparticle spectral weight as a function of the lattice parameter a in Fig.3.17 for pure NiS$_2$ at working pressures and in Fig.3.17 for NiS$_{2-x}$Se$_x$ at different Se–concentrations.

A close inspection of the Figure reveals a striking non–monotonic behavior of the spectral weight. For $a < 5.56$ Å (i.e. at the highest values of P), a nearly complete metalization has been reached. In this lattice region, the slow continuous gain of SW reducing a reflects the progressive enhancement of the kinetic energy (i.e. of the bandwidth W) due to the applied P. For $a > 5.56$ Å up to $a_{eq} \approx 5.68$ Å (namely the value corresponding to NiS$_2$ at ambient conditions), electronic correlation gets larger and the SW drops abruptly to zero as a consequence of the Mott transition. On further increasing a above a_{eq} by Se-alloying, the spectral weight (Fig. 3.17 starts to increase again, owing to the onset of the Se-MIT.

Despite the opposite behavior of the lattice parameter involved in the two MITs, a linear scaling relation ($x \approx 0.14$/GPa) has been formerly estab-

3.2 Optics of NiSe$_x$S$_{1-x}$

lished from low temperature DC *static* resistivity measurements. This scaling is based on the assumption of an equivalence between Se–alloying and P–application[129, 140]. However, the same scaling does not apply for the *dynamic* quantity, i.e., the optical SW, for which $x \approx 0.3/\text{GPa}$ seems to be more appropriated. Consequently there is *no* universal scaling and, hence, the concept of pressure doping equivalence appears inconsistent with the experimental data, just as in the case of V$_2$O$_3$.

ab initio LDA calculations The non–monotonic dependence of the SW on the lattice parameter as well as the difference in the SW transfer at low–energy suggest that the P and Se–MITs rely on two different microscopic mechanisms. To clarify these mechanisms, we performed self consistent TB–LMTO–LDA calculations [4], Nth order muffin-tin orbital (NMTO) downfolding [3, 6, 224], and the augmented plane waves plus local orbitals (APW+lo) technique within the framework of the Wien2K code [183]. Let us remark that recently other authors studied the NiS$_{2-x}$Se$_x$ compound with similar results [107].

Although the system belongs to the category of strongly correlated electron systems, and the LDA calculations cannot reproduce the charge transfer insulating behavior, we can gain valuable insight by analyzing the changes in the whole LDA spectrum (resolved in Ni and S/Se contributions) due to applied pressure and alloying.

In Fig. 3.18 we show in the upper part a sketch of the LDA DOS giving an overview of the general features and a definition of important quantities for the following discussion. In the lower four panels we show the actual LDA (Wien2k) results in an energy range ($\varepsilon_\text{F} \equiv 0\text{eV}$) from -6 to 3 eV corresponding to the dashed frame in the sketch. We show the partial DOS for the undoped NiS$_2$ at ambient pressure in the top left panel and the top right shows the calculated DOS for the NiS$_2$ structure at $P = 10\,\text{GPa}$. In the lower panels we show the NiSe$_2$ DOS calculated with the experimental structure as input on

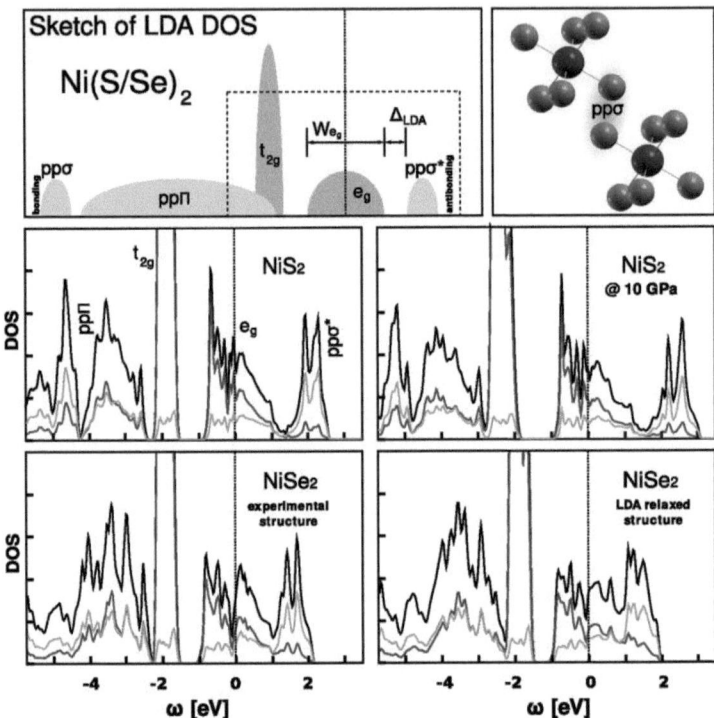

Fig. 3.18: Summary of Wien2k LDA calculations: In the top panel we show on the left hand side a sketch of the overall density of states for the $NiS_{2-x}Se_x$ system. For the framed region we show the actual LDA results in the four panels below with the Ni partial DOS plotted in dark gray and the S/Se partial DOS plotted in light gray. In the central panels we show the DOS for the undoped compound (left hand side) and the results for the structure at 10 GPa (right hand side). The bottom panels show the results for $NiSe_2$ calculated with the experimental structure (left hand side) and with the structure that was relaxed within a Wien2k LDA calculation (right hand side). From these results we can conclude that pressure and Se-doping modify the two fundamental parameters for the MIT, W_{e_g}/U and $W_{e_g}/\Delta_{\text{LDA}}$ in an orthogonal manner.

the left hand side and the LDA relaxed structure[15] on the right hand side[16]. In all four plots we find the following main features: At energies below -2.5 eV we find broad hybridized bands of Ni and S/Se p–states above which the very narrow Ni 3d t_{2g} states are located at ~ -2 eV. Around the Fermi energy we find the Ni 3d e_g states with a bandwidth of the order of $W_{e_g} \sim 2$ eV. The highest states we observe are *antibonding* S/Se $pp\sigma^*$ states associated to S–S or Se–Se pairs.

From the LDA spectra alone we can conclude that, on one hand the e_g bandwidth W_{e_g}, and on the other hand, the size of the gap between the e_g and the antibonding S/Se $pp\sigma^*$ states, Δ_{LDA}, are the significant quantities upon pressure application and Se doping respectively: For the NiS$_2$ pressure DOS (top right panel) we observe an increase in the bandwidth W of ≈ 0.5 eV for $P = 10\,\text{GPa}$ (i.e. a factor of $W_{e_g}^{10\text{GPa}} = 1.13 W_{e_g}$), while the the antibonding S–S/Se–Se $pp\sigma^*$ states are even *pushed up in energy* by ≈ 0.5 eV, i.e. the bandwidth–gap ratio $W_{e_g}/\Delta_{\text{LDA}}$ remains more or less constant. In strong contrast to this, the lattice expands in NiSe$_2$ (DOS in bottom left) due to the larger atomic radius of selenium. However, while we find changes of the e_g-bandwidth W_{e_g} to be negligible, the most striking effect is the reduction of Δ_{LDA} by ≈ 0.5 eV.

The pressure effect has a clear cut and intuitive understanding with a picture of increased overlaps of wave functions and a concomitant enhancement of the kinetic energy. The doping effect of the Δ_{LDA} reduction, however, has a less obvious reason. We can find the explanation for the smaller Δ_{LDA} actually in the distance of the S–S or Se–Se (see top right panel in figure 3.18) pair which is associated with the bonding antibonding $pp\sigma$–splitting. In NiS$_2$ the S–S pair distance is experimentally determined [62, 151] to be of the order of ~ 2.0 Å (the pressurized sample shows actually the same distance for the pair), while for NiSe$_2$ the Se–Se pair distance is ~ 2.4 Å. Hence, in NiSe$_2$ the pair hybridization and the associated splitting of bonding or antibonding states is *less* pronounced which, as a consequence, brings the $pp\sigma^*$ states closer to the e_g states. In this situation the Ni e_g – Se p hybridization

---

[15]Structural input was kindly provided by Jan Kunes (Academy of Sciences – Prague)
[16]The evident discrepancy between experimental and LDA calculated structure are discussed in the next paragraph

is stronger than the Ni e_g – S p hybridization which explains why the former compound is metallic. Hence, pressure and doping both enhance the kinetic energy of the system, but in very different ways.

Assuming now that the local interaction, expressed by a Hubbard U, *does not depend much on pressure or doping conditions*, theory and experiment allow us to conclude that applying pressure and Se–doping modify the two fundamental parameters for the MIT, W_{e_g}/U and W_{e_g}/Δ_{LDA} in an orthogonal manner: under pressure, W_{e_g}/Δ_{LDA} = const. and W_{e_g}/U increases, triggering the MIT. For Se–alloying on the other hand, W_{e_g} remains constant whereas the increase of W_{e_g}/Δ_{LDA} due to a closure of the Δ_{LDA} gap is responsible for the MIT. Hence, just like we observed for V_2O_3, the pressure–doping equivalence does not hold also in the case of the charge transfer insulator NiS_2.

effects of electronic correlations Let us close this section with the discussion of the effects caused by electronic correlations. The LDA results shown in Fig. 3.18 do not reproduce the insulating character of NiS_2. Further, the LDA DOS for $NiSe_2$ is not consistent with a correlated metallic state. In order to capture the correlation effects we derive a low energy model by means of NMTO downfolding and solve the resulting model with DMFT.

For the model we choose just the e_g–states around the Fermi energy so that the resulting Hamiltonian is a 2×2 matrix. The DMFT solution of this model is, in fact, able to capture the insulating nature of the NiS_2 ground state. However, the results for the e_g–only model are problematic concerning two aspects: i) First of all it is *by definition* not capable of capturing the charge transfer nature of the system and ii) the resulting ground state is "too insulating" in the sense that the MIT upon Se doping or pressure application cannot be comprehended. More specifically, the same model for the NiS_2 10GPa or the $NiSe_2$ remains insulating and bears no metallic solution in a physically reasonable parameter range. In Fig. 3.19 we show the calculated optical conductivity σ (main panel) and the corresponding spectral function $A(\omega)$ (inset) for the parameter set $U = 2.1$ eV, $V = 0.7$ eV, and $J = 0.7$ eV[17]. As we observe

---

[17]The value for J is already 0.2 eV below the (constrained LDA)estimated value, and a smaller choice of U would lead to an unphysical *attractive(!)* interaction $V - J$ between two e_g

3.2 Optics of NiSe$_x$S$_{1-x}$

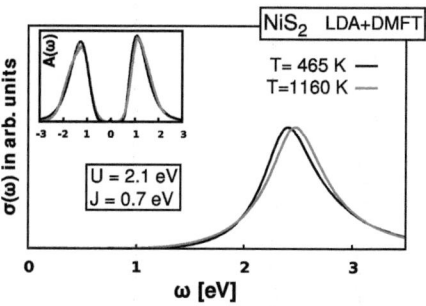

Fig. 3.19: LDA+DMFT result for the optical conductivity (main plot) and the spectral function (inset) of NiS$_2$. The spectra are clearly insulating with a very large gap ($\sim$ 1 eV for $A(\omega)$) although the values for U, V, and J are far below the lower threshold of the expected values. This discrepancy can be attributed to the disregard of the details of the p–degrees of freedom.

the spectra are clearly insulating with a very large gap ($\sim$ 1 eV for $A(\omega)$) although the values for U, V, and J are way below the lower threshold of the expected (see footnote 17).

The reason for both of the mentioned inconsistencies is the entanglement and interplay of the Ni 3d e_g–states and the S/Se 3p/4p states which is simply beyond a 2×2 e_g–only basis. Especially for the NiSe$_2$, where the gap Δ_{LDA} closes up completely we can intuitively understand, that the p–degrees of freedom should be taken into account explicitly. Moreover, the entanglement of the Se 4p states with the strongly correlated Ni 3d electrons might also explain the discrepancy between the experimentally measured structure: Comparing the two NiSe$_2$ cases in the bottom panels of Fig. 3.18 we find that the relaxed structure has an even smaller distance between the e_g– and the antibonding $pp\sigma^*$–states. This is a consequence of a smaller bonding antibonding Se $pp\sigma$–splitting, which corresponds to a larger Se–Se pair distance of $\sim$ 2.53 Å for the relaxed structure compared to $\sim$ 2.41 Å for the experimental one. Such discrepancy between an experimentally measured and a LDA calculated structure can indicate an important coupling between the correlated electronic structure and the crystal structure, which would need a full self–consistent *ab initio*+DMFT analysis as those we discuss in chapter 5.

conclusion In summary the NiS$_{2-x}$Se$_x$ system has turned to be more complex than previously assumed – similarly to the case of the Mott transition in

---

electrons with equal spin

3 Bulk 3d–transition metal compounds: Theory vs. Experiment

V_2O_3. Experimental data, together with results of *ab initio* LDA calculations have shown that the MIT upon pressure application and doping is driven by different microscopic mechanism. While the pressure induced MIT can be understood by a scenario of simple bandwidth broadening, the MIT induced by Se doping can only be described including a more involved d–p interplay which is tuned by the bonding antibonding splitting of the S–S / Se–Se pair states. A LDA +DMFT study showed that a simple 2×2 model only including the Ni 3d e_g–states is not sufficient to comprehend the MIT due to the disregard of the d–p interplay. Hence, similar to the case of V_2O_3, this archetype of a correlated CT insulator seems to display an even richer phase diagram and previously assumed equivalence of two pathways across a phase transition turned out to be a too simple picture.

4 Nickel oxide superstructures

In the previous chapter we have shown the application of the LDA+DMFT approach for correlated 3d–transition metal compounds including calculations for experimental observables like photoemission–, optical conductivity–, and x–ray absorption spectroscopy. Now, as promised by the title of this work, we leave the class of the "simple bulk" materials and turn to materials which incorporate a superstructure. These also include compounds that have been synthesized with the help of fast evolving experimental techniques. In such materials often the "effective dimensionality" is reduced and the interplay between low dimensionality and strong correlations often results in spectacular new physics. Quasi two–dimensional layer systems present indeed a variety of fascinating phenomena among which the most remarkable, without question, is the high temperature superconductivity. In this chapter we discuss the possibility of finding analogies between the high T_C cuprate superconductors and nickel oxide based compounds:

In the first section we review the cuprates and, specifically, introduce the key features which were the motivation for the search of cuprate physics in nickelates. In the second section we will present the results of our LDA+DMFT study on the LnSrNiO$_4$ (Ln=La,Nd,Eu) perovskites series with quite promising results. Finally, in the third section we will present a study on nickelate heterostructures which are synthesized with the help of extremely advanced experimental techniques like, e.g, molecular beam epitaxy and actually turn out to bear an even higher potential for cuprate analogies.

The band structure calculations and the NMTO downfolding for the low energy models of the nickelate compounds were performed by Xiaoping Yang in

4 Nickel oxide superstructures

the group of Prof. O.K. Andersen at the Max–Planck–Institute für Festkörperforschung Stuttgart, Germany.

4.1 Starting from the high T_C cuprates

The motivation for our work starts, as mentioned above, with the cuprates. The discovery of high temperature superconductivity (HTSC) in hole doped cuprates [15] initiated the quest for finding related transition metal oxides with comparable or even higher transition temperatures. In some systems such as ruthenates [118] and cobaltates [193] superconductivity has actually been found. However, in these t_{2g} systems superconductivity is very different from that in cuprates and transition temperatures (T_c's) are considerably lower. But which oxides, besides cuprates, are most promising for getting high T_c's? The basic band structure of the hole–doped cuprates is that of a single two–dimensional Cu $3d_{x^2-y^2}$–like band which is less than half–filled (configuration d^{9-h}). In this situation, antiferromagnetic fluctuations prevail and play a crucial role in many suggested scenarios for the superconductivity. The Fermi surface (FS) of this $x^2 - y^2$ band has been observed in many overdoped cuprates and found to agree with the predictions of density–functional (LDA) band theory. Yet, a thorough understanding of the mechanism of high T_C is still missing or, at least, the matter of an extremely controversial debate.

In the last twenty years many compounds with CuO_2 layers have been synthesized. All exhibit a phase diagram with T_C going through a maximum as a function of doping. A widely accepted explanation is, that at low doping superconductivity is destroyed with rising temperature by the loss of phase coherence, and at high doping by pair breaking [55]. For the materials dependence of T_C at optimal doping, i.e., $T_C^{max.}$, the only known, but not understood, systematics is that for materials with multiple CuO_2 layers, such as $HgBa_2Ca_{n-1}Cu_nO_{2n+2}$, $T_C^{max.}$ increases with the number of layers, n, until

4.1 Starting from the high T_C cuprates

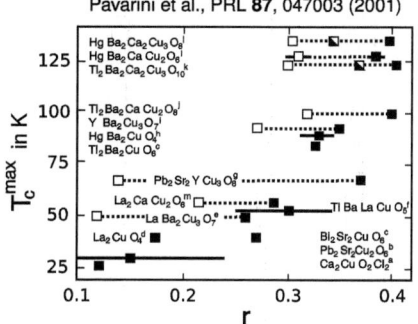

Fig. 4.1: Correlation between calculated range parameter r and observed critical temperature at optimal doping $T_C^{\text{max.}}$. Filled squares: single-layer materials or, for multi-layers r for the most bonding subband. Empty squares: most antibonding subband. Half–filled squares: nonbonding subband. Dotted lines connect subband values. Bars give k_z dispersion of r in primitive tetragonal materials.

$n = 3$.

There is also little clue as to why for n fixed, $T_C^{\text{max.}}$ depends strongly on the family, e.g., why for $n = 1$, $T_C^{\text{max.}}$ is 40K for La_2CuO_4 and 85K for $Tl_2Ba_2CuO_6$, although the Néel temperatures are fairly similar. A wealth of structural data has been obtained, and correlations between structure and $T_C^{\text{max.}}$ have often been looked for as functions of doping, pressure, uniaxial strain, and family. However, the large number of structural and compositional parameters makes it difficult to find what besides doping controls the superconductivity. In 2001 Pavarini et al. [154] suggested, that the material dependence enters the Hamiltonian of these compounds already in the one–electron part and should, hence, be captured by means of LDA. In fact, a careful analysis of the the band structure of the hole–doped cuprates by means of muffin tin orbitals (MTOs) uncovered a *dimensionless material dependent parameter* which is correlated to $T_C^{\text{max.}}$. Our work, which will be discussed in the subsequent sections, is motivated by this observation.

Two–orbital model and the *range* parameter Let us start by revisiting the work of Pavarini et al. [154] to set the stage for our analysis. The main conclusion drawn in this work is, that the basic model for understanding the materials trend in the cuprates is a two band model consisting of a *planar* $x^2 - y^2$ band around the Fermi energy and spread within the CuO plane and

4 Nickel oxide superstructures

an *axial* band perpendicular to it at higher energies[1]. The material dependent parameter is, in principle, the energy distance of the *axial* band to the *planar* band around the Fermi energy. This energy difference can be rewritten in a dimensionless *range* parameter r which becomes larger for a smaller energy difference ($r^{-1} \propto |\varepsilon_{\text{axial}} - \varepsilon_{\text{planar}}|$, see [154] for details). In multilayer systems the same trend is found by taking the lowest lying, i.e. most bonding, *axial* subband. In turn, this indicates that $T_C^{\text{max.}}$ increases with the number of layers because, as a consequence of the bonding/antibonding splitting due to the stacking, the *axial* band is lowered in energy. In Fig. 4.1 we show the plot of $T_C^{\text{max.}}$ for several compounds as a function of the *range* parameter (References for the compounds from "a" to "m" are [5, 221, 203, 30, 70, 39, 192, 191, 163, 16, 190, 204, 88, 31]). The filled squares in this plot correspond to the single–layer materials or the most bonding subband for multilayers. The empty and half filled squares denote the most antibonding subband and nonbonding subband, respectively.

To understand this remarkable observation better, let us take a look at the details of the *planar / axial* band scenario. LDA calculations for a large number of cuprate families have revealed that whereas the dispersion along the nodal direction (Z–A) in the Brillouin zone is always the same, the energy of the saddlepoints at $\left(\frac{\pi}{a}, 0\right)$ and $\left(0, \frac{\pi}{a}\right)$ depends on the material and is lower for materials with higher T_C^{max}. The reason for this correlation is not understood, but the reason for the change of band shape is clearly that the *planar* $x^2 - y^2$ orbital is hybridizing with a higher lying, material–dependent *axial* ($m = 0$) *orbital*. The energy of the *axial* orbital is of the order of 10eV and decreases for cuprates with increasing T_C^{max}. This axial orbital is essentially the antibonding linear combination of Cu $4s$ and apical O $2p_z$, so that its energy decreases if their interaction decreases, e.g., by increasing $z_{\text{Cu-O}}$. The composition of the *axial* orbital of the cuprates with respect to the Fermi energy is visualized in the energy level diagram shown in the upper right panel of Fig. 4.2. Also, as an example, we show a projected contour plot in the x/z plane of the

---

[1]Let us remark, that this two band model for the cuprates is *one* possibility of downfolding the full cuprate Hamiltonian. In chapter 5 we will elaborate more on this point, making also a connection to Emery–like models

4.1 Starting from the high T_C cuprates

hybridization between the *planar* and *axial* orbitals for the La_2CuO_4 system. As mentioned above we do not take the energy of the *axial* orbital as the specific parameter but instead use the dimensionless *range* parameter, r: The lower the energy of the *axial* orbital is, the higher $T_C^{max.}$ and the higher r. Let us remark, that we can account for this materials trend also if we still downfold further to a *single band model* which captures only the *planar* band around ε_F and, hence, the structure of the Fermi surface. Restricted to nearest and next nearest neighbor hoppings, such model can be written as:

$$\varepsilon(\mathbf{k}) = -2t\left(\cos(k_x) + \cos(k_y)\right) + 4t'\left(\cos(k_x) \cdot \cos(k_y)\right) \quad (4.1)$$

where t and t' are the nearest and next nearest neighbor hopping respectively. The effect of the *axial* band is encoded in the next nearest neighbor hopping t'. Hence, for the single band model (4.1) the materials trend can be expressed as the ratio of t' to t. In fact, for materials with low $T_C^{max.}$ ($< 50K$), which corresponds to a "small" value of r, the *range* parameter can be approximated *directly* as the ratio $r \approx t'/t$. The cuprates with the highest $T_C^{max.}$ (~ 140 K) have $r \sim 0.4$. If one could lower the energy of the axial orbital right down to the Fermi level, r would have the value $1/2$. In Fig. 4.3 we show the Fermi surfaces, at half filling, calculated with Eq. (4.1) for different values of t'/t

In order to give an actual example of the *planar/axial* band interplay we plot in Fig. 4.4 the corresponding bands for La_2CuO_4 (left hand side), with a range parameter of $r = 0.15$, and $HgBa_2CuO_4$ (right hand side), with a range parameter of $r > 0.3$. The color in the plots code the orbital character – from now on and throughout this entire chapter: light gray = *axial*, **black** = *planar*. Those two cases show, that the r–parameter can be understood, in a way, as a measure for the energy distance of the two (planar & axial) bands, i.e. $r^{-1} \propto |\varepsilon_{axial} - \varepsilon_{planar}|$. The observation of this empirical trend suggests the idea of a more general mechanism underlying the cuprates behavior. On one side this provides important hints for a better theoretical understanding of the cuprates superconductivity. In addition this trend also suggests the idea to find compounds other than the cuprates with similar characteristics

4 Nickel oxide superstructures

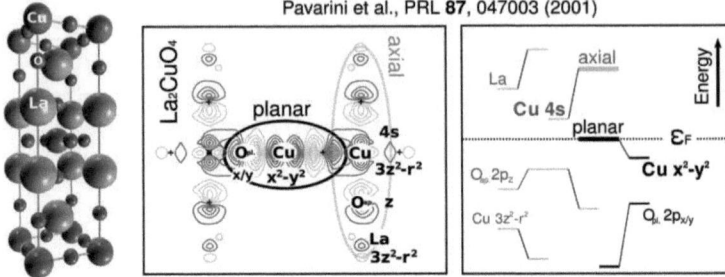

Fig. 4.2: Left hand side: Crystal structure of LaCuO$_4$. Center: Muffin Tin Orbital showing the interplay of the *planar* (black) and *axial* (light gray) states in La$_2$CuO$_4$. The plane is perpendicular to the layers. Right: schematic composition of the *planar* and *axial* states in terms of their coupling constituents.

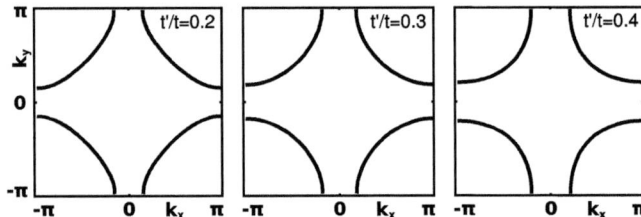

Fig. 4.3: Cross-section of the FS for the two dimensional single band model (4.1) including nearest t and next nearest t' neighbor hopping. From left to right we show the Fermi surface for $t'/t = 0.2, 0.3, 0.4$.

4.1 Starting from the high T_C cuprates

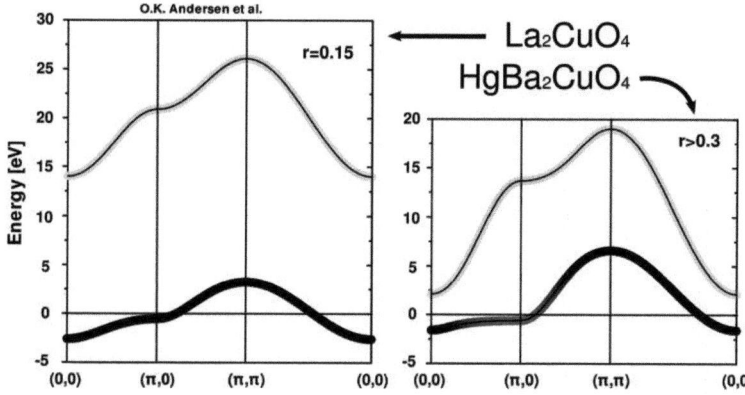

Fig. 4.4: Band structures of the *planar* (black)/ *axial* (light gray) – two band model for La_2CuO_4 (left) and $HgBa_2CuO_4$ (right). The closer the *axial* band is to the Fermi level, the higher $T_C^{max.}$ for the doped compound. Plots by O.K. Andersen *et al.*

in their electronic structure and as a consequence, hopefully, new routes for high T_C superconductivity.

Cu^{2+} vs. $Ni^{3+/2+}$ in tetragonal environments Our search for compounds of similar characteristics as the cuprates leads us one step to the left in the periodic table: Nickel. In order to understand how the *planar/axial* scenario could be "mapped" on nickel–based compounds, we should at first understand the electronic configuration of nickel compared to Copper.

In the cuprates, copper is found to be in a 2+ state with nine d–electrons in a tetragonal crystal field environment. Considering the analogous nickel–based compounds, i.e. the nickelates, we find stable Ni^{2+} (eight d–electrons) and Ni^{3+} (seven d–electrons) systems. The best way to understand a possible analogy between the two material classes is to draw a simple picture of crystal field one–electron states like we find it in a chemistry or solid state physics textbook. In Fig. 4.5 on the left hand side we draw the levels of Cu^{2+}, Ni^{3+}, and Ni^{2+} in a cubic environment with a perfect octahedral coordination

101

4 Nickel oxide superstructures

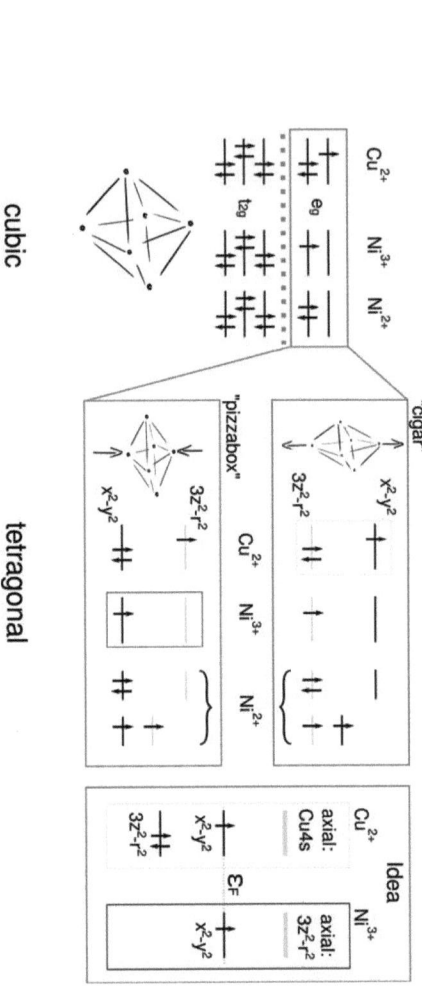

Fig. 4.5: Simple single electron energy level diagrams for Cu^{2+}, Ni^{3+}, and Ni^{2+}. Starting from the cubic state (left hand side) we show the splitting of the e_g manifold due to trigonal distortion (along the fourfold axis) of the ligand field. For Ni^{3+} we observe a half filled state at the Fermi energy ε_F of $x^2 - y^2$ or $3z^2 - r^2$ symmetry depending on the sign of the distortion. Ni^{2+} shows either two half filled states at ε_F or a full and an empty state. Hence, only the Ni^{3+} compound is a suitable candidate for the cuprate analouge scenario.

102

4.1 Starting from the high T_C cuprates

of ligands. The cubic eigenstates of the d–electrons are the lower lying t_{2g} states and the higher lying e_g states. In all three cases the t_{2g} states are completely filled and we will neglect them since we are interested only in the states around the Fermi energy. For Cu^{2+} there are three electrons in the e_g states, for Ni^{3+} and Ni^{2+} there are only one and two electrons respectively. Starting from the cubic case, the tetragonal situation can now be understood as a distortion of the octahedron along the four–fold axis, e.g., in the z–direction. This distortion can be done by either "pulling" on the top and bottom, which creates the so called "cigar"–shaped tetragonal distortion, or we can "compress" the octahedron creating a "pizza box". Both distortions lead to a lifting of degeneracy of the e_g states, but in opposite direction. The "cigar" distortion pushes the $x^2 - y^2$ states up in energy since now the ligands in the xy–plane towards which the $x^2 - y^2$ lobes point are closer compared to the ligands on the z–axis. On the other hand, the "pizza box" distortion increases the energy of the $3z^2 - r^2$ states since now the ligands on the z–axis are closer. If we now consider the filling again, we see that for the Cu^{2+} there remains only one electron in the higher lying e_g state while the lower one is completely filled. In the cuprates we find precisely this situation where the relevant, i.e., half-filled higher lying, orbital is the $x^2 - y^2$ which means we find the "cigar"-distorted case (framed light gray). The nickelates on the other hands have less electrons and, hence the lower one of the two e_g states is occupied. In the case of Ni^{2+} we either find completely full or empty bands when the crystal field splitting is larger compared to the Coulomb interaction or we find two bands at half filling. Neither of these situations is eligible to resemble the half filled cuprate situation. On the other hand, we find in the Ni^{3+} case always the *lower* e_g orbital filled with one electron. Hence, the $x^2 - y^2$ is at half-filling in the "pizza box"–distorted case (framed dark gray) with the $3z^2 - r^2$ at *higher* energies (and not full and low as in the cuprates). Now we can easily formulate the *planar/axial* orbital idea as we also sketch it in Fig. 4.5 on the right hand side: In the cuprates the half filled *planar* $x^2 - y^2$ states hybridize with the *axial* states consisting mainly of Cu 4s – In the nickelates this role of the *axial* states could be played by the $3z^2 - r^2$ states which are split of to *higher* energies. Depending on the size of the splitting we, i.e the energy of the *axial* $3z^2 - r^2$ we expect very large *range* parameters

4 Nickel oxide superstructures

like for the cuprates with the highest $T_C^{\text{max.}}$.

The first nickel oxides considered for the above described scenario were the LnSrNiO$_4$ perovskites (Ln=lanthanide element). However, as a consequence of negative results from from band structure calculations with LDA and LDA+U [8] they were quickly discarded as good candidates for superconductivity.
According to recent experimental findings though [207], the possibilities of this class of materials may have been underestimated. The new experimental insights were motivation enough to revisit the perovskites and perform LDA+DMFT calculations for a series of three compounds (R=La,Nd,Eu) in order to compare the results to the experiments. Our findings show, that the physics of these compounds have a much higher potential than thought in the past.
Moreover, besides the search for compounds that are "already on the market", new technologies and experimental techniques offer the possibility to encounter the quest of finding a high T$_C$ nickelate a little bit more "active": Growing heterostructures with molecular beam epitaxy on specific substrates and selected chemical constituents and similar synthesis techniques have evolved to maturity and bear great possibilities to control the "parameters" of a compound, not longer only in theoretical calculations, but in the real world. In the section after next we discuss as an example the LaAlO$_3$/LaNiO$_3$ 1/1 layered structure and conclude that the proposed idea of synthesizing nickel oxide heterostructures looks quite promising in order to find cuprate analogue systems.

4.2 LnSrNiO$_4$ perovskites

Even before the discovery of superconductivity in the La$_{2-x}$Ba$_x$CuO$_4$, other oxides crystallizing in the K$_2$NiF$_4$ structure were widely studied for the rich variety of their structural and physical properties. Among them, special attention has been paid to rare earth nickelates, Ln$_2$Ni$_4$O$_4$ ("Ln" = lanthanide

4.2 LnSrNiO$_4$ perovskites

element). In Fig. 4.6 on the lower left hand side we show a sketch of the LnSrNiO$_4$ structure. The main motive in the structure are tetragonally distorted NiO$_6$ octahedron which share their four in-plane corners. The hopping of the axial $3z^2 - r^2$-orbital is suppressed and, so the hope, it will play the role of the axial orbital higher in energy than the planar $x^2 - y^2$-orbital as sketched on the right hand side in Fig. 4.5.

In LDA+U calculations [8] the nickel e$_g$ states open a symmetry broken gap in a S=1 state and the lower Hubbard band shifts below the oxygen p-states which remained at the Fermi energy. The conclusion was that mobile O-holes at the Fermi level will be strongly antiferromagnetically coupled to the localized Ni spins and thus LaSrNiO$_4$ should be regarded as a heavy electron metal with a large Kondo temperature. However, recent experimental results [207] of ARPES suggest a d-like Fermi surface in these compounds. The discrepancy between the experimental data and the LDA+U calculations could be originated from the roughness of the static mean field assumption of LDA+U [181]. Hence, encouraged by the experimental findings, we performed LDA+DMFT calculations in the same fashion as they were done in the case of V$_2$O$_3$ (see chapter 2). We will start by describing the LDA results and the NMTO downfolding to the relevant (low energy) degrees of freedom to compare the La;Nd;Eu series. In the following paragraph we discuss the LDA+DMFT results and show spectral functions and the results for the local self energy as well as Fermi surfaces for the correlated cases.

LDA and NMTO downfolding for the LnSrNiO$_4$ series On the right hand side of Fig. 4.6 we show the NMTO (black/light gray) bands downfolded to the *planar* ($x^2 - y^2$: black) and *axial* ($3z^2 - r^2$: light gray) degrees of freedom plotted on top of the LDA bands (thin lines). These two bands are the antibonding Ni–O $pd\sigma$ e$_g$–bands. They lie well above the Ni–O $pd\pi$ t$_{2g}$–bands (the thin lines) and other p–degrees of freedom. Also above the e$_g$–bands we find a rather good separation from other (Note, for ESNO the f-bands have a too large hybridization with states around the Fermi energy so that a calculation which treats them simply as core states turned out to be unstable). Due to the separation of the e$_g$–bands, our downfolded NMTO two–band model should

4 Nickel oxide superstructures

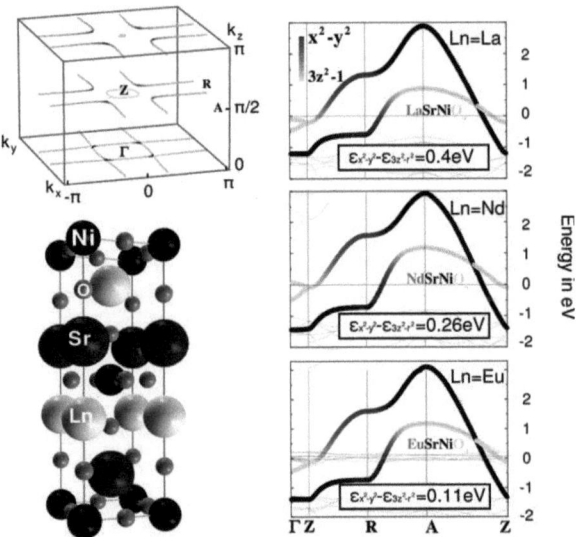

Fig. 4.6: Left hand side: Conventional unit cell of the LnSrNiO$_4$ series and above the NdSrNiO$_4$ Fermi surface plotted for $k_z = 0$, $\pi/2$, and π. The coloring gives the x^2-y^2 vs $3z^2-1$ e_g Wannier-function character. Right hand side: LDA band structure plots for three compounds of the series with Ln=La, Nd, Eu. The Bloch vector is along the lines $\Gamma(0,0,0) - Z\left(0,0,\frac{\pi}{c}\right) - R\left(0,\frac{\pi}{a},\frac{\pi}{c}\right) - A\left(\frac{\pi}{a},\frac{\pi}{a},\frac{\pi}{c}\right) - Z\left(0,0,\frac{\pi}{c}\right)$. Also in the band structure plots the color codes the orbital character.

be able to capture well the low energy excitations of the system.

In the LDA plots we can also observe a fundamental characteristic of the e_g orbitals. Namely, although locally the e_g orbitals are orthogonal eigenstates of the tetragonal crystal field they generally hybridize with one another in k–space. In real space this simply means, that the e_g–orbitals show non-local interorbital hopping. This k–dependent hopping has certain symmetry directions where the hybridization is zero. For example in the $k_x = k_y$–plane, i.e., the nodal direction, whereas in other directions it is maximal, as in the $k_x = \pi$ or k_y–plane. In this respect, e_g electrons behave very differently than t_{2g} electrons, which have *no* inter–orbital hopping on a square lattice. Turning back to the band structures within the series Ln=La,Nd,Eu they appear to be quite similar: the $3z^2 - r^2$–band is narrower than the $x^2 - y^2$–band as a consequence of the already mentioned suppression of the hopping along the z–axis. In all three cases the centers of gravity (COG) of the bands are quite close. The exact values of the energy differences of the COG difference are given in the plots. The $\varepsilon_{x^2-y^2} - \varepsilon_{3z^2-r^2}$ splitting increases when going from Eu to Nd to La due to the differnt crystal field potentials. As we will see the small changes of the splitting will be extremely important in the later LDA+DMFT calculation. As we can see, the $3z^2 - r^2$ band still crosses the Fermi energy and contributes significantly to the Fermi surface: On the upper left hand side of Fig. 4.6 we show the LDA Fermi surface for the NdSrNiO$_4$ compound in three different $\mathbf{k}_z = 0; \pi/2; \pi$ slices. Also this plot is color coded and shows the hybridized character of the LDA result. It can be seen nicely that the light gray $3z^2 - r^2$ part of the Fermi surface is much more dispersive in the z-direction as opposed to the black part which reflects the two–dimensionality of the $x^2 - y^2$–band.

LDA+DMFT results Now we turn to the correlation effects. In metallic multi-band systems it is expected that the correlation effects enhance level split-tings. Actually, we encountered this effect ourselves in the results for V_2O_3 (see chapter 2). In the case of the nickelate compounds, the correlation medi-

4 Nickel oxide superstructures

ated "effective crystal field splitting"[2] controls the size of the splitting between the *planar* and *axial* states (see Fig. 4.5) and hence the resulting *range* parameter r.

After carefull analysis of the LDA band structures, where we found that the conduction bands in the paramagnetic phase are well separated from all other bands, we can study the effects of Coulomb correlations in the nickelate heterostructures using the two–band Hubbard Hamiltonian analogous to the three–band Hamiltonian of V_2O_3 from chapter 2:

$$\hat{H} = \sum_{\mathbf{k},mm',\sigma} H^{\mathbf{k}}_{mm'} \hat{c}^{\mathbf{k}\dagger}_{m\sigma} \hat{c}^{\mathbf{k}}_{m'\sigma} + U \sum_{i,m} \hat{n}^{i}_{m\uparrow} \hat{n}^{i}_{m\downarrow}$$
$$+ \sum_{i,mm',\sigma\sigma'} (V - \delta_{\sigma\sigma'} J) \hat{n}^{i}_{m\sigma} \hat{n}^{i}_{m' \neq m \sigma'} \quad (4.2)$$

Here, the on–site Coulomb terms, namely the intra and inter–orbital Coulomb repulsions, U and $V = U - 2J$, as well as the Hund's exchange, J, have been added to the NMTO e_g Wannier–function Hamiltonian, $H^{\mathbf{k}}_{mm'}$. As usual in a LDA+DMFT analysis, neither the pair hopping term-corresponding to highly excited states - nor the spin-flip term - which poses a severe sign problem in the QMC simulation [80] - have been included[3]. We solve this Hubbard Hamiltonian for 1/4 (i.e. one electron on each Ni site) filling in the single–site DMFT approximation for the paramagnetic phase and at a high enough temperature ($1160\,\mathrm{K} = 0.1\,\mathrm{eV}/\mathrm{k_B}$) to use the Hirsch–Fye Quantum Monte Carlo method.

---

[2] Let us clarify two important issues concerning the terminology, in order to avoid confusion. i) First of all, the term *crystal–field* is historically connected to the electrostatic potential of the neighboring ligands. However, also in this work the expression does not only refer to the electrostatic part of the potential, but includes as well the hybridization with the ligands, i.e., charge exchange (hopping). Strictly speaking our crystal field is really a *ligand field*. ii) Second, the *effective crystal field splitting* is not related to the crystal/ligand field in its original meaning – it is a way to express the level splittings due to local correlation effects in terms of the one electron crystal field quantum numbers, i.e., the orbital labels – therefore "effective" CF.

[3] In our case this is partly justified by the fact that we have only one e_g per site on average.

4.2 LnSrNiO$_4$ perovskites

Summary of results The calculations have been performed for various values of the interaction parameter U and the results can be summarized in three statements:

1. Our DMFT calculations confirm that for a metallic multiband system, one effect of the Coulomb correlations is to enhance the splitting between the subbands while, on the other hand, the coherent excitations are quasiparticle renormalized.

2. Within the accuracy of our calculations the level splitting just enhances splittings that are already present on the LDA level but never reverses them. This means the enhanced splitting is in opposite direction as desired and we remain with a $3z^2 - r^2$ band and Fermi surface (like in the "cigar"–distorted case for Ni^{3+} shown in Fig. 4.5).

3. Due to the previous point, the DMFT results for these compounds are extremely sensitive to the starting LDA input and we find that especially the EuSrNiO$_4$ is in a way "on the edge". Specifically, for EuSrNiO$_4$ a small shift of 100 or 200meV of the starting LDA Hamiltonian leads to the cuprate like scenario which was described in the beginning, as we will see shortly.

Detailed discussion In Fig. 4.7 on the left hand side we show the real part of the LDA+DMFT self energy for all three compounds. On the right hand side we show the corresponding k–integrated spectral functions $A(\omega)$. The results shown were obtained with $U = 5.5$eV and $J = 0.7$eV. From the real part of the self energy we clearly see that the correlation mediated level splitting always lowers the $3z^2 - r^2$ states (light gray), while the $x^2 - y^2$ (black) states are shifted higher in energy and, hence, deplete their charge. In the spectral functions

4 Nickel oxide superstructures

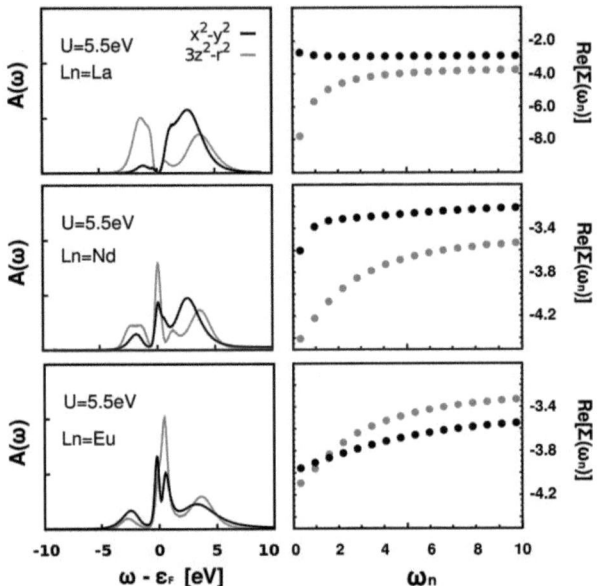

Fig. 4.7: LDA+DMFT results for the LnSrNiO$_4$ for Ln=La(top), Nd(center), and Eu(bottom). On the left hand side we show the spectral function (in arbitrary units) and on the right hand side the corresponding real part of the self energy as a function of imaginary Matsubara frequencies. The extrapolation of $\Re\Sigma(i\omega_n \to 0)$ leads to the crystal field enhancement caused by correlation effects. Note, for ESNO the high frequency part (i.e. the Hartree part of the DMFT self energy) $\Re\Sigma(i\omega_n \to \infty)$ the energy splitting has *opposite sign* compared to $\Re\Sigma(i\omega_n \to 0)$. See text for further discussion.

4.2 LnSrNiO$_4$ perovskites

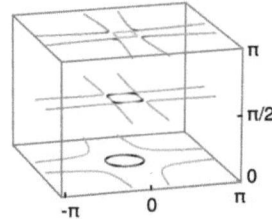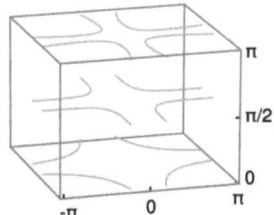

Fig. 4.8: LDA+DMFT Fermi surface of NdSrNiO$_4$ for $U = 5.5eV$ (left hand side) and $U = 5.8eV$ (right hand side). We can see that the "effective crystal field splitting" leads to a single sheet Fermi surface of mostly $3z^2 - r^2$ (light gray) character. The *axial* character manifests itself in the large k_z-dispersion of the Fermi surface.

this can be seen as the transfer of the $x^2 - y^2$ states above the Fermi energy. The consequences for the Fermi surface are visualized in Fig. 4.8 where we plot the Fermi surface for the NdSrNiO$_4$ case corresponding to the $U = 5.5eV$ spectrum and for an even more correlated, but still metallic, case $U = 5.8eV$. Hence, our system made a transition from a state with two FS sheets to a single, in this case $3z^2 - r^2$, FS sheet before reaching the insulating state. In the band–picture this means, that the bottom of the $x^2 - y^2$–band is pushed above the Fermi energy. This "sheet selective" transition can, in a way, be compared to the orbital selective Mott transition [11]. However, it should be stressed that, due to the strong inter orbital hybridization the e_g–systems in cubic, or close to cubic symmetry, are not expected to display true orbital selective transitions. For more detailed discussions concerning the evolution with the value of U we refer to the last subsection of this chapter, where we discuss model calculations of a hybridized two–band system at quarter filling.

Coming back to the main results and, specifically to Fig. 4.8 we can conclude that we do not find the cuprate analogous scenario but rather the "upside down"–version: An $x^2 - y^2$–band pushed above the Fermi energy becoming a kind of *band–insulator* at large U, whereas the $3z^2 - r^2$ resides as a strongly correlated conduction band before becoming a *Mott insulator* at large U.

4 Nickel oxide superstructures

Although the general trend is qualitatively the same for all three compounds, there is an important quantitative difference. We specifically choose to plot the spectra for $U = 5.5$eV, since it represents not only a value which might be a realistic choice, but more importantly, it is a choice of spectra which help to illustrate this difference: While LaSrNiO$_4$ is already an insulator for this value of U, NdSrNiO$_4$ still shows coherent quasi particle excitation around the Fermi energy as well as the Eu system which is even less correlated due to the fact that still *both* bands provide hopping channels (compare also the corresponding self energies). An important observation is, that these differences seem related to the LDA starting point of the onsite energies. Namely, if we compare the differences of the onsite energies in the LDA (shown in the band structure plots of Fig. 4.6) to the final level split in the LDA+DMFT spectral functions due to the real part of the self energy (Fig. 4.7), we assert that the DMFT always amplifies an effect which is visible in the LDA data already. We can conclude that the correlation effects taken into account on the DMFT level simply amplify the effects already present in the LDA band structure.

Moreover, for the ESNO system at $U = 5.5$ eV, we can observe an interresting detail. The onsite energy of the $z^2 - r^2$ states is in all compounds for the LnSrNiO$_4$ series *lower* than the onsite energy of the $x^2 - y^2$ states (for ESNO the difference is 100 meV). However, due to the very different bandwidths the filling of the $3z^2 - r^2$ band is *lower* that the filling of the $x^2 - y^2$ band. Nonetheless, the $x^2 - y^2$ states are pushed higher in energy by the correlation effects. This tells us, in turn, that the shift effect is beyond a static density density type of interaction. For ESNO at $U = 5.5$ eV, the static part has, in fact, the *opposite* sign compared to the effective crystal field enhancement, which is reflected in the high frequency behaviour of the real part of the self energy in Fig. 4.7. For the other compounds we find the same behaviour at smaller values of the interaction parameter.

After realizing the relation between initial LDA splitting and final LDA+DMFT slitting, let us have a closer look at the LDA numbers again. While in LaSrNiO$_4$ the $3z^2 - r^2$ lies ≈ 0.4eV above the $x^2 - y^2$, we find ≈ 0.25eV in NdSrNiO$_4$ and even less ≈ 0.1eV in EuSrNiO$_4$. While a 1meV accuracy might

4.2 LnSrNiO$_4$ perovskites

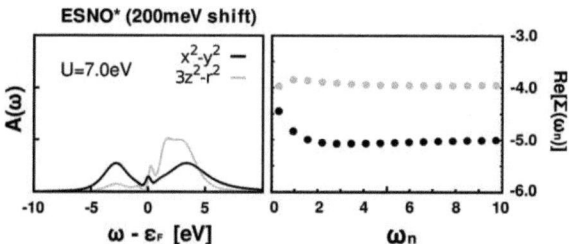

Fig. 4.9: LDA+DMFT results for a modified EuSrNiO$_4$ LDA Hamiltonian. The onsite energies of the $3z^2 - r^2$ states were "artificially" shifted by 200meV up in energy. The resulting correlation induced crystal field enhancement has reversed its sign compared to the calculation for the original LDA Hamiltonian

be usual for band structure calculations in non–correlated systems, an accuracy of 100meV in a correlated systems is usually already smaller than our error bars. A shift of 100meV might change the situation dramatically, if the hypothesis holds that DMFT correlation effects amplify LDA level splittings. In order to test this idea we performed additional calculations for the EuSrNiO$_4$ compound, in which we introduced an *artificial* shift of 100meV and 200meV of the *axial* $z^2 - r^2$ states in the LDA Hamiltonian which was used in the k–integrated Dyson equation of the DMFT self consistency. The result of these calculations clearly confirmed our idea. Even for the calculation with just 100meV shift the trend is reversed and the $z^2 - r^2$-states are pushed up in energy now leaving the $x^2 - y^2$-band at the Fermi energy. In Fig. 4.9 we show the spectral function and the self energy of the calculation with 200meV at a rather large U value of $U = 7.0$eV. This result indicates that the LnSrNiO$_4$ compounds – at least the EuSrNiO$_4$ is *on the edge* concerning the FS sheet selective transition. In fact, experimental ARPES results [207] indicate a $x^2 - y^2$–FS for the EuSrNiO$_4$ whereas the NdSrNiO$_4$ compound displays an ARPES spectrum which might be better described by a $3z^2 - r^2$ FS.

Let us close this section with a summary of the shifts between the two bands. In table 4.1 we show the comparison between the initial LDA onsite energy difference, and the respective difference in the real part of the

DMFT self–energies for all three compounds as well as for the artificially shifted EuSrNiO$_4$ case indicated by the asterix. For the values of $\Delta\Re\Sigma_{\text{DMFT}}$ we *did not* take the extrapolated values $\Re\Sigma_{\text{DMFT}}(i\omega \to 0)$ but the difference of the self energies at the smallest Matsubara frequency, i.e. $\Delta\Re\Sigma_{\text{DMFT}} = \Re\Sigma_{\text{DMFT}}^{3z^2-r^2}(i\omega_1) - \Re\Sigma_{\text{DMFT}}^{x^2-y^2}(i\omega_1)$. When we inspect the plots in Fig. 4.8 closely we realize that the value of our $\Re\Sigma_{\text{DMFT}}$ presents, in fact, a lower boundary for the real shift.

It should be stressed, that the artificial shift we introduced in this last case was not meant to be a realistic shift. It was performed to (in the end successfully) proof the hypothesis of the instability of these systems[4]. Such a shift, which corresponds to the shift of the *local* on site energy of the states seemed the most natural parameter for a single site DMFT. Non–local corrections which are beyond the single site DMFT might be sensitive to k–dependent details of the band structure and, hence, yield different results in the vicinity of the instability, i.e., a $x^2 - y^2$–FS sheet, directly from the LDA Hamiltonian. In order to capture also non–local effects, extensions of the DMFT have been proposed. For short range correlations the cellular DMFT or the Dynamical Cluster Approximation (DCA) are employed which are based on solving small clusters in real and k–space respectively, whereas the diagrammatic approaches such as the DΓA [206, 100] or the Dual Fermion [174] scheme are capable also to include the effect of long range spatial correlations.

---

[4] Such instability is not unusual for strongly correlated systems: the MIT in V_2O_3 can be reproduced within LDA+DMFT taking into account a change of the lattice parameters of less than 4% [14]

Tab. 4.1: Summary of shifts

	$\Delta\varepsilon_{\text{LDA}}$	$\Delta\Re\Sigma_{\text{DMFT}}$
LaSrNiO$_4$	0.40 eV	−5.10 eV
NdSrNiO$_4$	0.26 eV	−0.80 eV
EuSrNiO$_4$	0.11 eV	−0.14 eV
EuSrNiO$_4^*$	−0.09 eV	+0.48 eV at $U = 7.0$ eV

4.3 LaAlO$_3$/LaNiO$_3$

*The main part of the following discussion is published in the APS Journal "Physical Review Letters" [75]: PRL **103**, 016401, (2009)*

Let us review the "story so far". In the previous section the LnSrNiO$_4$ compounds were discussed with the motivation of finding nickelate systems which may display cuprate analog physics including, most importantly, superconductivity at a high critical temperatures. Following the planar/axial orbital scenario of the work of Pavarini et al. [154], the part of the axial orbital in the Ni^{3+} nickelates should be played by the $3z^2 - r^2$ band. Further, we saw that correlation effects, treated within the LDA+DMFT approach, split the Ni e_g states resulting in an "effective crystal field splitting". The class of the LnSrNiO$_4$ (especially the EuSrNiO$_4$) turned out to be "on the edge" and, from the calculation perspective, we found a delicate relation between the LDA+DMFT results and the spectrum of the one electron, i.e., LDA part of the Hamiltonian. The sensitivity of the system to these details allows for the question wether we can influence the system from the "outside" by means of changing external parameters like, e.g., pressure. These thoughts can be taken even one step further: As sophisticated experimental techniques to grow transition–metal oxides heterostructures are nowadays available and mature, our quest got a new direction: Novel effectively two–dimensional sys-

4 Nickel oxide superstructures

tems could be engineered to suit our needs.

Recently the following idea to get to a cuprate–like situation in nickelates was presented [34]: Sandwiching a LaNiO$_3$ layer between layers of an insulating oxide such as LaAlO$_3$ will confine the $3z^2 - 1$ orbital in the z-direction and may remove this band from the Fermi level, thus leaving the one Ni 3d electron in the $x^2 - y^2$ band. The difference to the "bulk" nickelates we described in the previous section is that, in the case of an "artificial" heterostructure, *we* have the choice of many parameters in our own hands, such as the possibility of using different substrates for growing the structure and performing chemical substitution of the specific chemical constituents. Indeed, a major reconstruction of orbital states at oxide interfaces recently has been observed [33], and this kind of phenomenon could lead to novel phases not present in the bulk. Extensive theoretical studies of mechanisms for orbital selection in correlated systems (see e.g. [11]) have revealed the complexity of this problem, where details of the electronic structure and lattice distortions play decisive roles. It is therefore crucial to examine nickelate heterostructures by means of state–of–the–art theoretical methods and find the optimal conditions for $x^2 - y^2$ orbital selection. In this section we present results of electronic–structure calculations using LDA+DMFT.

Summary of results The results can be understood in complete analogy to the previous section. We find that the comparably large value of the hopping between the x^2-y^2 and $3z^2-r^2$ orbitals substantially reduces the effects of correlations in the $3z^2 - r^2$ orbital. Nevertheless, we do find that the correlations are able to shift the bottom of the hybridizing e_g bands sufficiently relative to each other, i.e., the effective crystal field splitting, to yield a FS with only *one* sheet. In the case of the heterostructure which we will focus on, this remaining sheet has predominantly $x^2 - y^2$ character and a shape like in the cuprates with the highest $T_{c\,\text{max}}$ [154], even showing a more extreme curvature. Moreover we will extend the discussion analyzing how the above mentioned FS–transition can be practically driven in the growth process. As we will see, this is possible by stretching the in–plane lattice constants by suit-

4.3 LaAlO$_3$/LaNiO$_3$

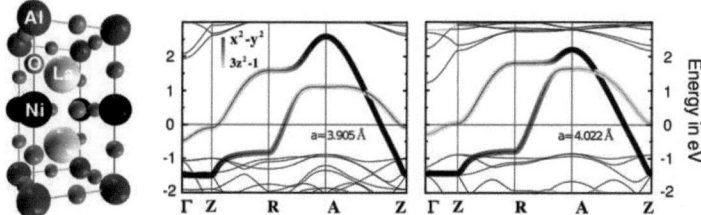

Fig. 4.10: The 1/1 LaNiO$_3$/LaAlO$_3$ heterostructure (left hand side) and its LDA (NMTO) bandstructures without (center) and with (right hand side) strain. The Bloch vector is along the lines $\Gamma(0,0,0) - Z(0,0,\frac{\pi}{c}) - R(0,\frac{\pi}{a},\frac{\pi}{c}) - A(\frac{\pi}{a},\frac{\pi}{a},\frac{\pi}{c}) - Z(0,0,\frac{\pi}{c})$. The shading gives the $x^2 - y^2$ vs $3z^2 - 1$ e_g Wannier-function character.

able choice of substrate in order to reduce the correlation–strength needed to produce a single–sheet FS.

Finally, a toy model tailored for this compound revealed its tendency towards strong antiferromagnetic fluctuations, even larger than in the cuprates. Our conclusion is, that nickelate heterostructures hold the basic ingredients for high-temperature superconductivity.

LDA and NMTO downfolding for LaNiO$_3$/LaAlO$_3$ Here we give results for the simplest, 1/1 superlattice LaNiO$_3$/LaAlO$_3$ = LaO-NiO$_2$-LaO-AlO$_2$ shown in the left hand side of Fig. 4.10. For the in–plane lattice constant a we first took that of SrTiO$_3$ often used as substrate, whereby the Ni–O and Al–O distance in the x– and y–directions became: $x_{\text{Ni-O}} = 1.95$ Å, not far from the value in pseudo–cubic LaNiO$_3$. The lattice constant c we took as the sum of those of pseudo–cubic LaNiO$_3$ and LaAlO$_3$; afterwards the position of apical O was relaxed within the LDA [106] to yield $z_{\text{Ni-O}} = 1.91$ Å, i.e., a 2% smaller value than $x_{\text{Ni-O}}$. Next, we expanded the LaNiO$_3$/LaAlO$_3$ heterostructure in the x- and y-directions by 3%, as might be achieved by growing LaNiO$_3$/LaAlO$_3$ on a PrScO$_3$ substrate, to yield $x_{\text{Ni-O}} = 2.01$ Å. With the concomitant 6% contraction in the z-direction, relaxation of the apical-oxygen position within the LDA finally lead to $z_{\text{Ni-O}} = 1.81$ Å.

4 Nickel oxide superstructures

Fig. 4.10 shows the LDA energy bands for the two differently strained heterostructures in a 5-eV region around the Fermi level ($\equiv 0$). The two solid bands are, as in the previous section, the 1/4-full Ni-O $pd\sigma$ antibonding e_g bands, which are pushed up above the less antibonding Ni-O $pd\pi$ t_{2g} bands (thin bands). The later are lying below -1 eV and well above the Ni–O, Al–O, and La–O bonding bands below the frame of the figure. The antibonding Al–O and La–O bands (thin bands above 1-2 eV) lie respectively 9 and ~ 5 eV above their bonding counterparts, and as a result there is a comfortable 2-3 eV gap above the top of the antibonding t_{2g} bands in which the two antibonding e_g bands reside.

Let us compare the situation to the LnSrNiO$_4$ compounds which we discussed in the previous section. The coloring of the e_g bands again gives the relative $x^2 - y^2$ and $3z^2 - 1$ characters in the NMTO ($N = 2$) Wannier-function representation. Concerning the symmetry of the k-dependent inter e_g-hoppings we find the same situation as before. Also the k_z- dispersion of the heterostructures is comparably low as in the LnSrNiO$_4$ series. However, we observe even with the naked eye, that now the $3z^2 - r^2$-states are higher in energy with respect to the $x^2 - y^2$-states. In fact, even for the unstrained compound (left band structure plot in Fig. 4.10) the COG of the $3z^2 - r^2$-band lies 150 meV *above* the $x^2 - y^2$-bands COG. The reason for this difference to the LnSrNiO$_4$ series is the stronger apical $pd\sigma$ hybridization along $z_{\text{Ni-O}}$ in the LaAlO$_3$/LaNiO$_3$ compound which, besides the energy shift, also enlarges the bandwidth of the $3z^2 - r^2$-band as can be seen comparing the band structure plots in Fig. 4.6 and Fig. 4.10. Thus, we can reasonably expect that in our DMFT calculations the *axial* $3z^2 - r^2$-states this time will be pushed higher in energy, leaving a correlated $x^2 - y^2$-conduction band and Fermi surface. As we by now understand, the COG energy difference of the bands gives the direction of the splitting. The details of the FS sheet selective transition depend, of course, on the specific band structure features:

The bottoms of both bands are along ΓZ,, i.e., for $k_x=k_y=0$. That of the $x^2 - y^2$ band is at $-1.5\,\text{eV}$ and does not disperse with k_z, while that of the $3z^2 - r^2$ band is at -0.5 eV at Γ and disperses upwards to -0.1 eV at Z. The bottom of the $3z^2 - r^2$ band is thus 1 eV $\approx 1/4$ e_g bandwidth above that of the $x^2 - y^2$ band. Straining by 3% is seen to shift the bottom of the $3z^2 - r^2$ band up by further 0.2 eV. The LDA FS thus has *two* sheets, and reducing it to one would require moving the $3z^2-r^2$ band above the x^2-y^2 band at Γ by an additional 0.5 eV for the unstrained and by an additional 0.3 eV for the strained superlattice. That the x^2-y^2 Wannier orbital is more populated than $3z^2-r^2$ (in LDA the ratio is 70/30 for the unstrained superlattice) is mainly due to the *confinement* in the z-direction. Consider for simplicity the dispersions in the $k_x = \pm k_y \equiv k$ planes where the $3z^2-r^2$ and x^2-y^2 orbitals do not hybridize: In cubic bulk LaNiO$_3$, $\varepsilon_{3z^2-r^2}(k, k_z) \approx -\cos ak - 2\cos ak_z$, with respect to the center of the e_g band and in units of the k_z-hopping amplitude $|t_{dd\sigma}|$ of the *axial* $3z^2-r^2$, while $\varepsilon_{x^2-y^2}(k,k_z) \approx -3\cos ak$ independently of k_z because the k_z-hopping amplitude $t_{dd\delta}$ of the *planar* x^2-y^2 is negligible. This means that both bands extend from $-3|t_{dd\sigma}|$ at $(0,0,0)$ to $+3|t_{dd\sigma}|$ at $\left(\frac{\pi}{a}, \frac{\pi}{a}, \frac{\pi}{a}\right)$ in the bulk. Substituting now every second LaNiO$_3$ layer by an "insulating" LaAlO$_3$ layer, forces the Bloch waves to have nodes approximately at the AlO$_2$ planes, so that only waves with $k_z < \frac{\pi}{c} \sim \frac{\pi}{2a}$ are allowed. Together with the dispersion relation $\varepsilon_{3z^2-r^2}(k, k_z) \approx -\cos ak - 2\cos ak_z$, we can see that this restriction causes the bottom of the $3z^2-r^2$ band to be pushed up by $\sim 2|t_{dd\sigma}|$, i.e., by $\sim 1/3$ of the e_g bandwidth. The exact position of the nodes, and hence the upwards shift of the $3z^2-r^2$ band, depends on the scattering properties of the insulating layer. This suggests that the band structure can be tuned by choice of the insulating layer.

Turning back and comparing the present case to the undoped (d^9) cuprates, the LDA bandstructures are roughly similar to this. But, in the case of the cuprates, the $3z^2-r^2$ band does *not* play the role of the *axial* orbital. It is full and lies in the region of the t_{2g} bands (compare Fig. 4.5). Filling this cuprate $3z^2-r^2$ band has annihilated the $pd\sigma$ bond to apical oxygen and thereby caused the distance between Cu and the apical oxygen $z_{\text{Cu-O}}$ to increase well beyond the Cu in-plane oxygen distance $x_{\text{Cu-O}}$ (i.e. the "cigar"-distorted case depicted in Fig. 4.5), whereby the antibonding push-up of the $3z^2-r^2$ band has been lost. The half-full $pd\sigma$ antibonding x^2-y^2 band lies a bit lower with respect to the O and cation bands than in the nickelates because the position of the $3d$-level in Cu is lower than in Ni.

Now, if we compare the shape of the conduction band of our nickelate het-

4 Nickel oxide superstructures

erostructures to the cuprate conduction band we find a resembalance in particular to the conduction band of the cuprates with the highest $T_{c\,\text{max}}$. Hence, the idea described earlier of applying the *planar/axial* scenario seems feasible for the nickelates *if* the *axial* (i.e. the $3z^2 - r^2$ for the nickelates) states can be removed completely from the Fermi energy and pushed up in energy. Since the bands of the nickelate $3z^2 - r^2$ states cross ε_F for the LDA bands shown in Fig. 4.10, they have $r > 1/2$. Therefore, engineering these heterostructures should presumably first aim at reducing r towards that of the cuprates with the highest $T_{c\,\text{max}}$ ($r{\sim}0.4$), by moving the energy of the $3z^2 - r^2$ band at Γ, i.e., the bottom of the band, well above ε_F. This requires increasing the interaction (the *ligand field*) between Ni $3d_{3z^2-r^2}$ and apical O $2p_z$, e.g., by reducing $z_{\text{Ni-O}}$, or in other words making a "pizzabox"–distortion. As we shall see, electronic correlations help a lot in that respect. However this does not necessarily lead to HTSC, because although the same value of r gives the same band shape for nickelates and cuprates, their conduction–band Wannier orbitals are not completely identical.

LDA+DMFT results In order to perform the DMFT calculations we employ the same Hamiltonian (4.2) as for the LnSrNiO$_4$ series with the NMTO e_g Hamiltonian, $H^{\mathbf{k}}_{mm'}$. We solve this Hubbard Hamiltonian for the same condition as in the LnSrNiO$_4$ series, i.e., 1 electron per Ni site and $\beta = 0.1\text{eV}^{-1}$.

The results of our DMFT calculations can be understood straightforwardly in the light of the discussion of the previous section. Besides the reduction of the spectral weight of the coherent excitations around the Fermi energy, i.e., the renormalization of the quasi particle mass, we again find that electronic correlations determins an enhancement of the "effective crystal field splitting". Specifically, for the undoped superlattice with $J=0.7\,\text{eV}$ and U increasing, we find that the bottom of the $3z^2 - r^2$ band is driven up and passes the Fermi level when U exceeds 6.4 eV for the unstrained and 5.7 eV for the strained structure. Hereafter the FS has only one sheet, a large $\left(\frac{\pi}{a}, \frac{\pi}{a}\right)$-centered hole cylinder whose shape can be seen from Fig. 4.11 to be similar to that found in the cuprates with the highest $T_{c\,\text{max}}$. It is of course possible that the strong nesting of this FS makes it unstable with respect to spin or/and charge-density waves with $q_x \sim \frac{\pi}{2a}$ and $q_y \sim \frac{\pi}{2a}$, similar to what has been found in cuprates. At the point where the second FS sheet disappears, $r=1/2$ and the ratio between the $x^2 - y^2$ and $3z^2 - r^2$ populations has increased to 80/20 for the unstrained – and even beyond that for the strained – superlattice. For slightly larger U, the upper quasiparticle band still exists above the Fermi level. This may be seen from the $3z^2 - r^2$ projected k-integrated spectral function on the right-hand side of Fig. 4.11. Reasonable changes of J slightly influence details of the Hubbard subbands, but not the physics of the transition.

Only when U reaches 7.4 eV for the unstrained and 6.5 eV for the strained superlattice, the lower quasiparticle band undergoes a Mott-transition, at which point the peak seen at the Fermi level in the right-hand side of Fig. 4.11 finally disappears (compare spectra for different U values in Fig. 4.13). For comparison, a half–filled cuprate band undergoes a Mott transition in DMFT for a critical value of U which increases with r and takes the value 4.5eV for $r=0.4$ [46]. This behavior for the cuprates is thus in line with what we find for the nickelate heterostructures where r_{LDA} (unstrained) $> r_{\text{LDA}}$ (strained) $\sim 1/2$. This supports our hope that the nickelates can be engineered such that, like in the cuprates, hole–doping in the proximity of the Mott transition can produce superconductivity. For nickelates there is even the possibility of engineering the e_g bands such that the real value of U falls between the one needed to reduce the FS to a single sheet and the one needed to eliminate this sheet by a Mott transition. If this can be achieved, superconductivity in the nickelates may occur even without doping. This is a remarkable result.

Toy model and magnetic ordering Finally, we need to estimate the strength of antiferromagnetic correlations, which are an important ingredient in several suggested scenarios for the high temperature cuprate superconductivity. Since our LDA+DMFT calculations would be prohibitively expensive for the study of low temperature magnetic properties, we extract essential information by merely diagonalizing the two–site version of the Hubbard Hamiltonian (4.2) obtained by Fourier transformation of $H_{mm'}^{\mathbf{k}}$ and a truncation to a diatomic molecule directed along x. The complete results, as well as a detailed

4 Nickel oxide superstructures

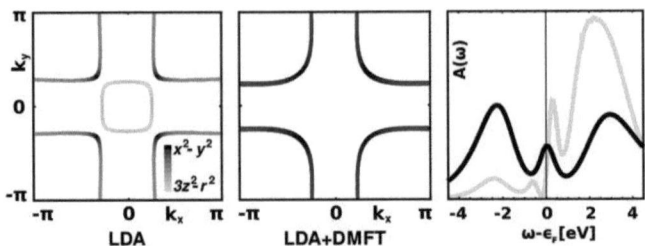

Fig. 4.11: Cross-section of the FS with the $k_z=0$ plane for the unstrained 1/1 heterostructure; left: LDA, middle: LDA+DMFT (U=6.7 eV). Right: LDA+DMFT k–integrated spectral functions (in arbitrary units) projected onto the x^2-y^2 (full) and $3z^2-r^2$ (dashed) Wannier functions.

discussion can be found in the last paragraph of the next section. Let us at this point just shortly comment that the ground state is always a spin singlet (see the energy level diagram in Fig. 4.19). Increasing the "effective crystal field splitting" Δ from 0 to ∞ leads to a demixing such that the orbital configuration changes from a mixture to pure x^2-y^2. For the LDA value, the orbital character is already predominantly x^2-y^2. From the distance between the singlet ground state and the triplet first excited state, we estimate the magnitude of the antiferromagnetic coupling constant to be $J_{AF} \sim 0.2\,\text{eV}$, i.e., somewhat higher than in cuprates.

In summary, our analysis of the 1/1 LaNiO$_3$/LaAlO$_3$ system shows that heterostructuring of d^7 nickelates is very promising since their physics contains the main ingredients of high–temperature superconductivity. In particular, we find that electronic correlations reduce the FS to a *single* sheet whose shape is similar to the one in the hole–doped cuprates with the highest $T_{c\,\text{max}}$. This sheet has a predominant *planar* x^2-y^2 character but also mixed in *axial* $3z^2-r^2$ character. Beside the substrate–induced strain, which we have discussed here, LDA calculations of similar nickelate based heterostructures have shown, that the choice of the other chemical constituents can sensitively influence the low energy physics of the system. In comparison to the LnSrNiO$_4$ structures of the preceeding section the nickelate heterostruc-

122

4.3 LaAlO$_3$/LaNiO$_3$

tures offer a larger spectrum of options to experimentally turn the parameters of the compound in order to find the high T$_C$–behaviour analogue to the cuprates.

4.3.1 MIT in a quarter–filled two–band system

In the preceeding sections we discussed two classes of nickel oxide systems in order to find cuprate analogous physics. Along the lines of the work of Pavarini *et el.* [154] we used the hybridizing *planar*–orbital / *axial*–orbital scenario performing LDA+DMFT calculations with the promising result that correlated nickel oxide superstructures bear the basic ingredients for high temperature superconductivity analogous to that of the cuprates.

Introduction Let us now shift the focus to a more general analysis of the model and the DMFT results for different kinds of interaction and hopping parameters. As it was described, for our DMFT we employ a two–band model obtained from NMTO downfolding on the e_g degrees of freedom, and performed the calculations assuming quarter filling, i.e., one electron in the two Ni e_g orbitals. In solids the cubic environment, or an environment easily derivable from the cubic one is extremely common. Examples are the tetragonal like environment in the cuprates (see Fig. 4.5), or the trigonal one in V$_2$O$_3$ (distortion along the C^3–axis of the octahedra). Depending on the specific crystal field splitting filling, either t_{2g} states (as in V$_2$O$_3$) or e_g states (as in the nickelates and cuprates) play the most important role in determining the low energy excitations of the system. Hence, a general study of models which include the basic features of such systems is desiarable. Within model calculations, half filled single band systems have been studied extensively by means of DMFT (see reviews [67, 81, 68, 105] and references therein). Besides that, also multi–band models were studied [11, 132, 21, 171, 104, 59, 49] and features like orbital selective transitions [7] were analysed. Yet, very often these multi–band studies assumed approximations, such as neglecting hy-

4 Nickel oxide superstructures

bridization, which prohibited to fully capture the rich physics driven by the orbital degrees of freedom in the correlated transition metal compounds.

Hence, we take the opportunity and discuss the features of our two–band systems at qurter filling in more detail. We will start with a brief review of the model we employ after deriving it from the LDA results. The first results we discuss are the spetral functions and the self energies for different values of the interaction parameters, followed by the double occupancy. Clearly the double occupancy is a more complicated quantity in the case of a multiband system compared to the single band case.

Next, we turn to the comparison of the model derived from the LDA calculations for a confined quasi two–dimensional system and a more homogeneous 3D model. Finally we discuss in more detail the issue of a magnetic order at low temperatures like we mentioned very briefly in the last part of the previous section. Further we remark that, parallel to ours, a work by Poteryaev *et al.*, in which similar models are discussed, was published [157]. Our work can be understood as complementary since we extend the study to the comparison between confined and homogeneous systems motivated by the specific role of the crystal geometry in the heterostructures.

Effective two-band model For our model study we start from the LMTO band structure for LaAlO$_3$/LaNiO$_3$ heterostructures grown on SrTiO$_3$. However, instead of employing NMTO to get a numerical Hamiltonian in k–space, we take the hopping elements in real space and truncate all hopping elements which are longer in range than the next nearest neighbour hopping. For our purposes of model calculations we prefer this simpler picture of the systems kinetics in real space and pay the price of a slightly less accurate fit of the LMTO bands.

In the basis $|x^2 - y^2\rangle = (1,0)$ and $|3z^2 - r^2\rangle = (0,1)$ the dispersion relation in k–space reads:

$$\varepsilon_{\mathbf{k}}^{2D} = -2(\cos(k_x) + \cos(k_y)) \cdot \begin{pmatrix} 0.45 & 0 \\ 0 & 0.17 \end{pmatrix}$$

$$+ 2(\cos(k_x) - \cos(k_y)) \cdot \begin{pmatrix} 0 & 0.28 \\ 0.28 & 0 \end{pmatrix}$$

$$- 4(\cos(k_x) \cdot \cos(k_y)) \cdot \begin{pmatrix} 0.09 & 0 \\ 0 & 0.03 \end{pmatrix} + \begin{pmatrix} 0 & 0 \\ 0 & 0.15 \end{pmatrix} \quad (4.3)$$

here the first and second summand account for the diagonal hopping and the off–diagonal hopping, i.e., the hybridization respectively, while the third summand presents the hopping to next nearest neighbours. The last term takes into account the onsite level splitting, that means a crystal field term. In real space, the Hamiltonian of our effective model reads:

$$\hat{H}_{\text{mod.}} = \sum_{iljm,\sigma} t_{iljm} c_{il\sigma}^\dagger c_{jm\sigma} + \hat{H}_{\text{int.}}, \quad (4.4)$$

where $l = (1,2)$ labels the two different orbitals, i.e., the *planar* $x^2 - y^2$ orbital and the *axial* $3z^2 - r^2$ orbital respectively. The hopping amplitudes t_{iljm} are the matrices in Eq. (4.3) with l, m as the orbital index and i, j as site index of nearest or next nearest neighbors on the cubic lattice.

The effects of the hybridization terms are clearly visible in the non–interacting density of states resulting from the noninteracting part of Hamiltonian (4.4) which is shown in Fig. 4.12. A comparison to the case without hybridization ($t_{12} = 0$) plotted as a thin line shows a big difference especially for the $3z^2 - r^2$-band. Its bandwidth is strongly enhanced if the hopping to the more mobile $x^2 - y^2$-orbitals is properly taken into account. This effect is important since otherwise the reduced hopping in the z-direction would more severely reduce the itinerancy of electrons in the $3z^2 - r^2$ orbital[5]. The Coulomb in-

---

[5] In the case of our $LaAlO_3/LaNiO_4$ heterostructures the $3z^2 - r^2$ orbital points towards the insulating AlO layers. Since these are insulating the hopping in z–direction is strongly reduced and can approximately be neglected.

4 Nickel oxide superstructures

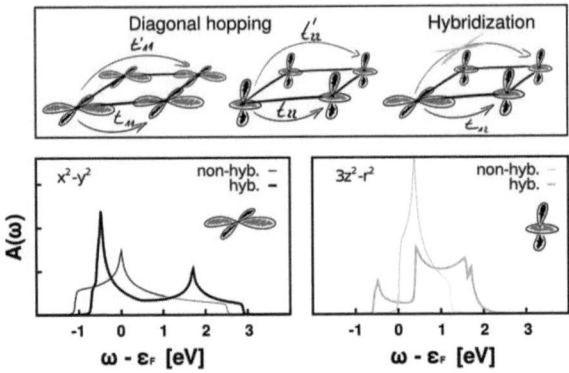

Fig. 4.12: Noninteracting DOS of the two e_g bands with (thick line) and without (thin line) $x^2 - y^2$-to-$3z^2 - r^2$ hybridization. The hybridization strongly enhances the $3z^2 - r^2$ bandwidth, leading to a remarkable itinerancy for this band despite the quasi two-dimensional nature of the LaNiO$_3$/LaAlO$_3$ heterostructure.

teraction terms are the same as in Hamiltonian (4.2) and for the solution of the impurity problem in the DMFT self consistency we use, as before, the quantum Monte Carlo simulations of Hirsch and Fye [85]. Afterwards the maximum entropy method (MEM: see chapter 2) is employed to perform the analytic continuation of the Green function to the real axis. Concerning the choice of the interaction parameter we perform a systematic analysis where U is varied in a regime of physically meaningful values in order to investigate the Mott–transition and its specific features. The Hund's exchange term, on the other hand, has been fixed to $J = 0.7$eV, a reasonable value close to that of atomic Ni. Assuming the deviation from cubic symmetry to be negligible for the Coulomb interaction parameters, we employ $V = U - 2J$.

Spectral Function and Self Energy Fig. 4.13 shows the evolution of the k–integrated spectral functions $A_l(\omega)$ with increasing U. Let us first focus on the $x^2 - y^2$ band (black curve). Already at $U = 4.4$ eV ($V = 3.0$ eV; first panel), the spectrum is strongly renormalized with respect to the non–interacting case:

4.3 LaAlO$_3$/LaNiO$_3$

A narrow quasiparticle peak at ε_F and Hubbard bands which are separated by an energy of the order of U are visible. As the value of the interaction parameter is increased to $U = 5.4$ eV and $U = 6.4$ eV (second and third panels), one observes a stronger renormalization of the quasiparticle peak and more pronounced Hubbard bands. A qualitative change is occurring for $U = 7.4$ eV (fourth panel) where the spectral weight at the Fermi level has vanished. That is, a Mott-Hubbard gap has formed, indicating the occurrence of the Mott metal–insulator transition between $U = 6.4$ and 7.4 eV.

The metal insulator transition of the $3z^2 - r^2$-band (grey curve) occurs for the same critical interaction of $U \gtrsim 6.4$ eV. It is, however, qualitatively different. When the interaction is increased one still observes a progressive reduction of the spectral weight at the Fermi level which has entirely disappeared for $U = 7.4$ eV. However, in this case this is due to a shift of the whole $3z^2 - r^2$ band above the Fermi level, highlighting the different nature of the metal–insulator transition. The shift with U is clearly seen for the quasiparticle peak. The remaining spectral weight below the Fermi energy, corresponding to the lower Hubbard band, is entirely due to the hybridization effect with the lower $x^2 - y^2$ Hubbard band.

The different nature of the metal–insulator transition in the $x^2 - y^2$- and the $3z^2 - r^2$-band emerges even more clearly from the analysis of the self energy. In Fig. 4.14, we show the evolution of the imaginary part of the self energy $\text{Im}\Sigma_{x^2-y^2}(\omega_n)$ with increasing interaction. Again, for the $x^2 - y^2$ band, the expected behavior of a standard Mott transition is found: The appearance of the insulating gap results from the divergence of $\text{Im}\Sigma_{x^2-y^2}(\omega)$ in the limit of zero frequency and zero temperature. In this limit the quasiparticle renormalization factor $Z = (1 - \frac{\partial \text{Im}\Sigma(0)}{\partial \omega_n})^{-1}$ goes to zero as a further hallmark of the Mott character of the transition in the $x^2 - y^2$-band (see black circles in the left panel of Fig.4.15). At the same time, however, the insulating transition of the $3z^2 - r^2$-band is not associated to any qualitative change in $\text{Im}\Sigma_{3z^2-r^2}(\omega)$. For all interactions a "metallic"–like bending can be observed in $\text{Im}\Sigma_{3z^2-r^2}(\omega)$ at low frequencies. As a consequence of this, the renormalization factor Z stays always finite (light gray crosses in the left panel of Fig.4.15). The disappearance of spectral weight at the Fermi energy in this case is determined

4 Nickel oxide superstructures

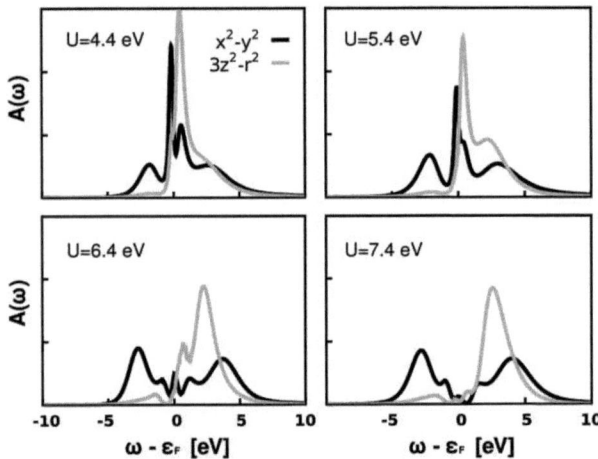

Fig. 4.13: Evolution of the LDA+DMFT spectrum with increasing Coulomb interaction $U = V - 2J$. The black curve shows the $x^2 - y^2$ spectrum and the light gray one the $3z^2 - r^2$ spectrum.

4.3 $LaAlO_3/LaNiO_3$

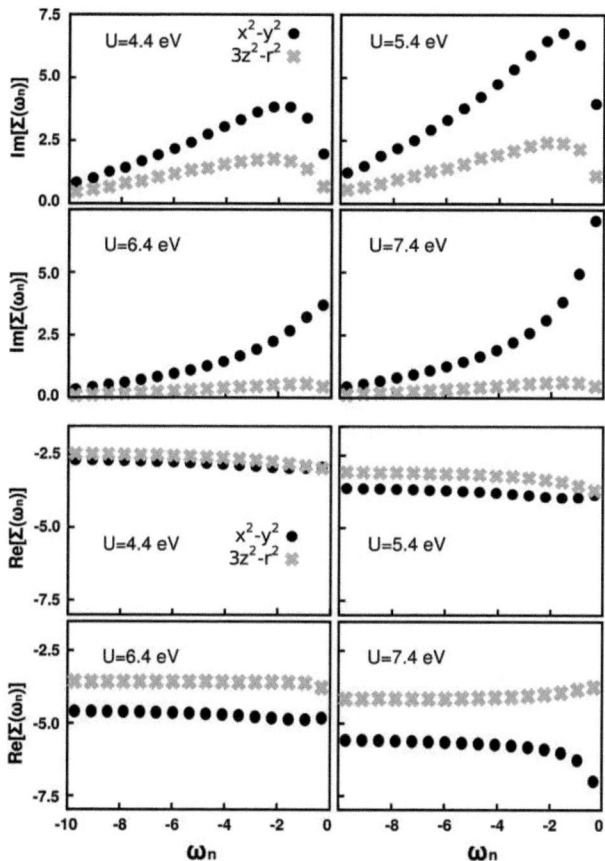

Fig. 4.14: Evolution of the LDA+DMFT self energy (vs. Matsubara frequencies) with increasing U. The black curves show the $x^2 - y^2$ and the light gray one the $3z^2 - r^2$ self energy. Via extrapolation of the imaginary and real part of the self energy to $\omega_n \to 0$ we can obtain the quasiparticle renormalization Z and the "crystal field" enhancment, respectively

4 Nickel oxide superstructures

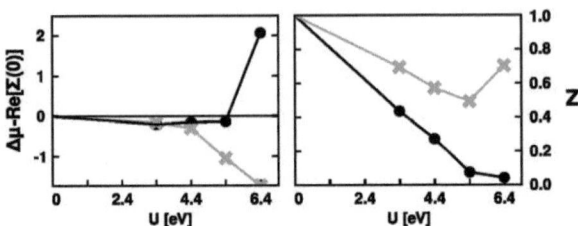

Fig. 4.15: Left hand side: Effective chemical potential $\mu - \mathrm{Re}\Sigma(0)$ relative to the noninteracting chemical potential for the $x^2 - y^2$ (black) and $3z^2 - r^2$ (light gray) orbital. At the Mott transition, the low frequency effective chemical potential of the d $3z^2 - r^2$ is strongly reduced so that this orbital is shifted above the Fermi energy, i.e., the $3z^2 - r^2$ orbital becomes "band"-insulating. Right hand side: Quasiparticle weight Z for the $x^2 - y^2$ (black) and $3z^2 - r^2$ (light gray) orbital. For the $x^2 - y^2$ orbital, we find $Z \to 0$, i.e., the (central) quasiparticle peak disappears and the $x^2 - y^2$ orbital becomes Mott-insulating.

by a large relative shift of the orbitals induced by interaction. This shift $\Delta_{CF}^{\mathrm{eff}}$, is a consequence of the strong variation of the real part of the self energies at zero frequency shown in the left panel of Fig. 4.15. More specifically, while the effective chemical potential of the $x^2 - y^2$-band (defined as $\mu - \mathrm{Re}\Sigma_{x^2-y^2}(0)$) stays always close to the noninteracting value up to the Mott transition, its value for the $3z^2 - r^2$-band is clearly decreasing with increasing V. As a result, the metal insulator transition of the $3z^2 - r^2$-band occurs for a value of $U \gtrsim 6.5$ eV, when the overall weight of this band is shifted above the Fermi energy.

A key feature in the physics we have observed here is the interplay between the two orbitals stemming from the fairly large hybridization terms in equation (4.3). An evident sign of such interplay is the concomitance of the metal insulator transition in both orbitals. To understand the underlying physics it is instructive to consider the different possible scenarios as well as the case without hybridization between the two bands.

The effects of increasing the interaction in partially filled bands is twofold: Firstly the on site double occupancy becomes more expensive and secondly the two bands are shifted with respect to each other. This gives rise to three possible transition scenarios. (i) If the shift between the bands is increasing slowly with interaction at a certain point the on-site double occupation becomes practically forbidden while both bands are still partially filled. As a result a simultaneous Mott transition would occur with $Z \to 0$ for both bands. If instead the shift is more rapidly increasing with interaction at a certain point one of the two bands will become empty leaving the other one at half filling. The nature of the transition would then depend on the value of the interaction at which the shifted band becomes empty. There are two possibilities: (ii) If the interaction is strong enough to forbid double occupancy in the filled band, this band will become abruptly insulating. Such a situation is reflected in a simultaneous – though qualitatively different – metal–insulator transition for both bands as we observe in Fig. 4.14. (iii) The other possibility is that the interaction is not yet strong enough to hinder the double occupations in the half filled band. The Mott transition in this band – differently from our case – would then not take place simultaneously with the depletion of the other band. This would be the only situation in which the metal–insulator–transition could be described in terms of an effective single band model.

The initial lifting of degeneracy (i.e., the crystal field splitting of the LDA calculation) is expected to push the system in the direction of situation (ii), which we observed for our model with an initial crystal field splitting of $\varepsilon_{3z^2-r^2} - \varepsilon_{x^2-y^2} = 150$ meV. In section 4.2, in the discussion of the LnSrNiO$_4$ series we saw that, to a first approximation, the sign of this splitting determines which of the bands will be pushed higher in energy.

Finally, we will discuss the role played by the hybridization, which has mainly two effects. The first one, which can be understood intuitively, is an effective broadening of the bandwidth principally for the $3z^2 - r^2$-band (see Fig. 4.12 for the noninteracting case) which leads to a more metallic behavior. The second effect is more intrinsic in the sense that the hybridization makes the bands more similar, obviously pushing the system towards situation (i). It is important to notice that both effects work against the crystal field splitting. As mentioned above the fingerprints of the hybridization can clearly be seen also in Fig. 4.13: Although the $3z^2 - r^2$–band is shifted above the Fermi energy at the metal insulator transition some residual spectral weight remains in the region of the lower $x^2 - y^2$-Hubbard band.

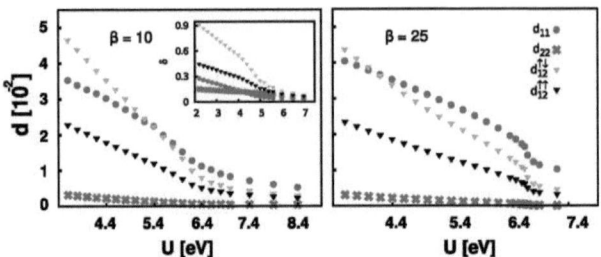

Fig. 4.16: Double occupancy vs. Coulomb interaction V. Shown are the double occupation in the $x^2 - y^2$ orbital (d_{11}), in the $3z^2 - r^2$ orbital (d_{22}) as well as the double occupation of both orbitals with parallel ($d_{12}^{\uparrow\uparrow}$) and antiparallel spin ($d_{12}^{\uparrow\downarrow}$). Inset: double occupancies normalized by the respective density.

Double occupancy As in the case of the half-filled Hubbard model, complementary information about the metal insulator transition can be extracted from the analysis of the double occupancy d, which is the derivative of the Helmholtz free energy H with respect to the interaction parameter U and can be interpreted as a sort of order parameter of the Mott metal–insulator transition[6]. However, such an analysis for a two-band model is more complicated since it involves not only the orbital-diagonal double occupations ($d_{mm} = \langle n_{m\uparrow} n_{m\downarrow} \rangle$) but also the orbital-offdiagonal parts with parallel ($d_{12}^{\uparrow\uparrow} = \langle n_{1\uparrow} n_{2\uparrow} \rangle$) and antiparallel spin orientations ($d_{12}^{\uparrow\bar{\sigma}} = \langle n_{1\uparrow} n_{2\downarrow} \rangle$).

The evolution of these four quantities is shown in Fig. 4.16 for two different temperatures. In both cases we observe the same trend for all double occupations which are decreasing with increasing interaction. Starting the analysis with the $x^2 - y^2$-orbital, which undergoes the Mott-Hubbard transition we observe that its double occupation displays a rapid decrease around $V = 5$ eV, i.e., close to the critical values of the interaction we can estimate form the spectral function in Fig. 4.13. At $\beta = 10$ (Fig. 4.16 left panel), however, the double occupancy behavior is thermally smeared out so that it is very difficult to obtain a quantitative estimate of the critical interaction value.

---

[6] However, since there is no symmetry breaking involved in the transition it is no *real* order parameter.

The Mott-Hubbard transition occurs in form of a very smooth crossover. A better estimate can be made when going down in temperature to $\beta = 25$ (Fig. 4.16 right panel), where we can see a very steep drop of d_{11} marking the transition point at $V \approx 5.5$ eV where we also found $Z \to 0$ for the $x^2 - y^2$ band. At a first glance the behavior of the other three double occupancies $d_{22}, d_{12}^{\uparrow\downarrow}$, and $d_{12}^{\uparrow\uparrow}$ appear to be more difficult to interpret: One could have expected that $d_{12}^{\uparrow\uparrow}$ and $d_{12}^{\uparrow\downarrow}$ should be the largest ones because they are energetically less expensive (V,V') in comparison with the d_{mm} (U). What we observe, however, is a crossing of $d_{12}^{\uparrow\uparrow}$ with d_{11} at around $V \approx 4$ eV and $V \approx 2.8$ eV for $\beta = 10$ and $\beta = 25$ respectively. Moreover, $d_{12}^{\uparrow\downarrow}$ is lower than d_{11} for all interaction values. This situation can be properly understood considering also the depletion of the $3z^2 - r^2$-band: The low values of d_{22}, $d_{12}^{\uparrow\downarrow}$, and $d_{12}^{\uparrow\uparrow}$ reflect the low electron density in the $3z^2 - r^2$-band. In order to disentangle these effects we have plotted the double occupancies normalized by the respective density in the inset of the left panel of Fig. 4.16 ($\delta_{mm} = d_{mm}/\langle n_m \rangle^2$, $\delta_{12}^{\uparrow\downarrow} = d_{12}^{\uparrow\downarrow}/(\langle n_{1\uparrow}\rangle\langle n_{2\downarrow}\rangle)$, and $\delta_{12}^{\uparrow\uparrow} = d_{12}^{\uparrow\uparrow}/(\langle n_{1\uparrow}\rangle\langle n_{2\uparrow}\rangle)$). In this way the hierarchy of the energetic consideration is restored and also a more similar behavior of all renormalized double occupancies occurs. On the other hand, the crossing we observe in the unrenormalized data can be interpreted as a clear signal that we are far from the situation (i) described above, that is a simultaneous Mott-Hubbard transition in both bands.

From the layered- 2D to the homogeneous 3D model After having discussed in detail the results for the quasi two–dimensional model, which was derived from LDA calculations for the LaAlO$_3$/LaNiO$_3$, we devote this paragraph to the comparison of our 2D–model to a more homogeneous, i.e., isotropic one. Specifically that means, we no longer restrict the axial $3z^2 - r^2$-orbital in the xy–plane, but allow for hopping along the z–axis adding the term

$$-2 \cdot (t_z = 0.6) \cos(k_z) \cdot \begin{pmatrix} 0 & 0 \\ 0 & 1 \end{pmatrix}$$

4 Nickel oxide superstructures

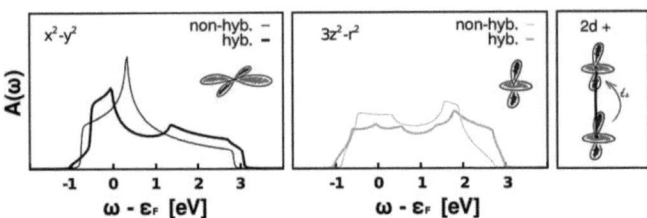

Fig. 4.17: Noninteracting DOS of the x^2-y^2 band (left) and the $3z^2-r^2$ band with (thick line) and without (thin line) x^2-y^2-to-$3z^2-r^2$ hybridization for the homogeneous system. The $3z^2-r^2$ band is even broader than in the layered case due to the additional mobility in the z direction.

to the kinetic part of our Hamiltonian (4.3). Such a situation could be realized for example in the non–layered structure LaNiO$_3$. The value for $t_z = 0.6$ in our calculation has been estimated using the table of Slater and Koster [188] for overlap integrals assuming a cubic symmetry. It is thus not to be seen as a true LDA+DMFT analysis of LaNiO$_3$ – the link to this compound is solely a motivation for the analysis of the modified model.

The noninteracting density of states for the homogeneous model is shown in Fig. 4.17. The main difference to the model of the layered system (see Fig. 4.12) is obviously, due to the additional mobility along the z direction, a much broader $3z^2-r^2$-band. As a result of this broadening the two different bands are much more similar in the homogeneous case. They are still not identical because the in–plane hopping matrix has been kept constant, i.e., equal to the LDA results for the heterostructure which is not cubic.

In Fig. 4.18 we show the spectral functions and self energies of the homogeneous system. On the left hand (right hand) side the data for a metallic (insulating) solution are plotted. They refer to $V = 6.0$ and $V = 8.0$ respectively. This indicates that the metal insulator transition occurs for larger values of the interaction than in the layered system as could have been expected. Another difference to the previous 2D case is the enhanced spectral weight of the $3z^2-r^2$-band below the Fermi energy. However, there is no qualitative dif-

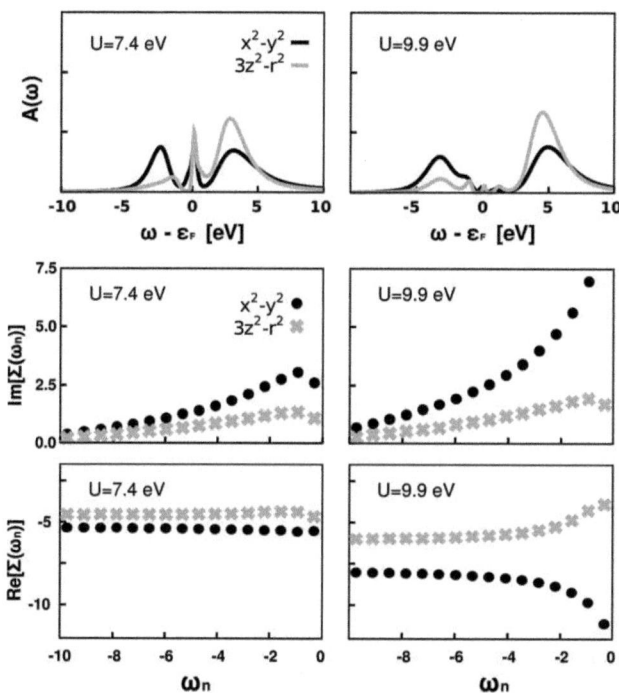

Fig. 4.18: Evolution of the spectrum and the LDA+DMFT self energy (vs. Matsubara frequencies) with increasing Coulomb interaction $U = V - 2J$ for the homogeneous case. The black curve shows the x^2-y^2 spectrum and the light gray one the $3z^2-r^2$ spectrum.

4 Nickel oxide superstructures

ference between the two bands in the transitions as is clearly demonstrated by the behavior of the imaginary part of the selfenergy shown in the bottom panels of Fig. 4.18: The evolution of the self energies qualitatively resemble the one shown in Fig. 4.14 displaying the same "metallic" down–bending. The only (quantitative) difference to be noticed w.r.t. the layered case is a larger imaginary part of the $3z^2 - r^2$-selfenergy at low-frequencies. This can be understood as a natural consequence of the closer resemblance of the bands in the noninteracting density of states. In terms of our precedent discussion we thus observe a situation slightly less far from the case (i) of a simultaneous double Mott transition for both bands than for the layered model.

Predominating spin fluctuations The model calculations discussed in the previous sections were performed for the paramagnetic phase. However, one expects the system to display magnetic and/or orbital order at sufficiently low temperatures. The treatment of such complex ordered phases is particularly difficult in LDA+DMFT. An insight into the physics can be obtained however by diagonalizing a "two-site version" of Hamiltonian 4.4 (with open boundary conditions). The analysis of the energy levels allows to infer the relevant fluctuations which dominate the low temperature physics of the system. On the left-hand side of Fig. 4.19 we are showing the calculated energy level diagrams as a function of U for the same set of parameters we have used for the LDA+DMFT calculation discussed in the previous section.

For the values of U considered, the mixing of the lowest energy levels with the energy costly double occupied states is very small, i.e., these states are already above the energy window which we plot in Fig. 4.19. More interestingly our results show that the ground state of the two site system is always a spin singlet with the two electrons residing predominantly in the same orbital ($x^2 - y^2$) on each site. This indicates an instability of the system at low temperatures towards a spin antiferro- and orbital ferro-ordered state. We have also studied the effect of varying the initial level splitting for a reasonable value of $U = 4.5$eV (right-hand side of Fig.4.19). The main effect of an enhancement of the crystal field splitting is to strengthen the tendency towards orbital ordering (of the x^2-y^2 and the $3z^2$-r^2 states) eventually leading

4.3 LaAlO$_3$/LaNiO$_3$

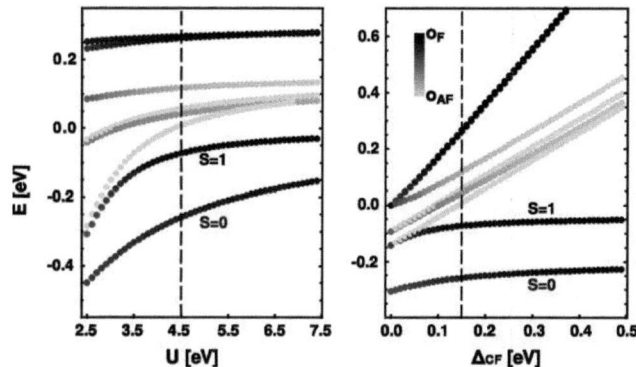

Fig. 4.19: Left hand side: Energy level diagram of the two site model as a function of the Coulomb repulsion U for $\Delta_{CF} = 0.15$eV (dashed line in the plot on the right-hand side). Right hand side: Same diagram as a function of Δ_{CF} for $U = 4.5$eV (dashed line in the plot on the left-hand side). The color of the points indicates the character of orbital ordering (o_F: orbital ferro; o_{AF}: orbital antiferro)

to purely one-band physics, with only the x^2-y^2 orbital involved.

A first analysis of the magnetic and orbital ordered phases in the La$_2$NiAlO$_6$ was carried out in Ref. [34] considering the following superexchange Hamiltonian, defined for a bond $ij \parallel \gamma$ in the NiO$_2$ plane:

$$H_{ij}^{(\gamma)} = (R_{\sigma,\pm,\pm} + R_{\sigma,\pm,\pm}^{CT})(\frac{1}{2} \pm \hat{\tau}_i^{(\gamma)})(\frac{1}{2} \pm \hat{\tau}_j^{(\gamma)})\hat{P}_{\sigma,ij} \quad (4.5)$$

with an implied sum over $\sigma = 0, 1$ and all combinations of $\pm, \pm$. $\hat{P}_{\sigma,ij}$ represents the projector to a singlet and a triplet state of two Ni^{3+} $S = \frac{1}{2}$ spins while $\frac{1}{2} \pm \hat{\tau}_j^{(\gamma)}$ selects the planar orbital in the plane perpendicular to the γ axis and the directional orbital along this axis.

Neglecting the charge transfer part of Hamiltonian (4.5) allows for a comparison with our two site calculation, albeit only for $\Delta_{CF} = 0$eV. In this way we can perform the first numerical estimate of the $R_{\sigma,\pm,\pm}$ coefficients of the superexchange Hamiltonian (4.5) as a function of the parameter of the microscopic Hamiltonian. The result of this comparison are shown in the left panel

4 Nickel oxide superstructures

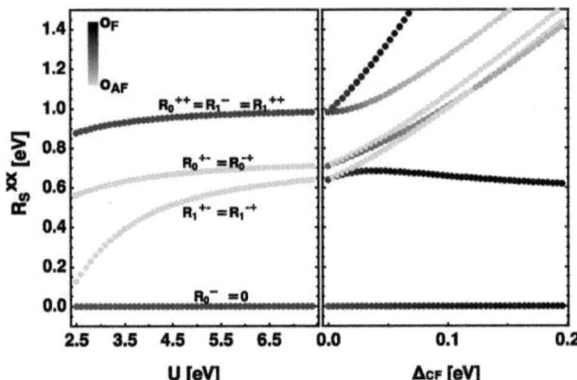

Fig. 4.20: Left hand side: Evaluation of the R-coefficients of Hamiltonian (4.5) as a function of the interaction parameter U calculated with the two site model in the $x^2 - y^2$ and $3z^2 - r^2$ basis for $\Delta_{CF} = 0$eV. Right hand side: Evolution with increasing crystal field splitting starting from the largest U value of the left panel ($U = 7.5$eV). The color of the points indicates the character of orbital ordering with the same scale as in Fig. 4.19.

of Fig. 4.20, where for simplicity we assumed a perfectly cubic symmetry for the overlap integrals. It is important to note here, that the orbital projectors in Eq. (4.5) do not refer to the $x^2 - y^2$ and $3z^2 - r^2$ basis we have used so far since the γ axis is lying in the NiO$_2$ plane. This is also reflected in the mixed color of the R^0_{--} coefficient which corresponds to the singlet ground state.

Finally, in the right panel we show the effect of turning on the crystal field splitting Δ_{CF}. The changes in the color clearly indicates a rapid demixing of the eigenstates, since the $x^2 - y^2$ and $3z^2 - r^2$ orbitals are eigenfunctions of the crystal field operator. This limits the applicability of Hamiltonian (4.5) to the cases of small crystal field splittings.

As for the magnetic fluctuations, the Kugel–Khomskii–type Hamiltonian (4.5) favors an antiferromagnetic spin–alignment along each pair. Depending on the crystal–field splitting, the preferred orbital orientation is along the bond or –for larger crystal field– a ferro–orbital occupation of the $x^2 - y^2$ orbital.

5 Expanding the basis set

Differently from the previous chapters, this last one focuses more on the conceptual part of the LDA+DMFT calculations.
As we will see within this chapter, the treatment of some correlated systems by means of LDA+DMFT cannot be done satisfyingly by employing a simple DMFT scheme we have described in chapter 2 and used in chapters 3 and 4. Hence, it became necessary to implement extensions of the standard procedures for the self consistent calculation. Here we present both the technical aspects as well as the results for realistic material calculations.
Specifically we will discuss extensions of the standard DMFT basis set for the low energy models and the connected definition of the interaction parameters. In the first section we present the so called LDA+HartreeDMFT approach.
The basis set of a model Hamiltonian is extended to include ligands of the local DMFT "impurity" site. Within this section we will present the new scheme pedagogically by taking the Ni based heterostructures from chapter 4 as an example of an application. Afterwards we turn to the high T_C cuprates and analyze Emery–like models, which incorporate one d–band and two p–bands. It will be shown that for these models both the extension of the basis set as well as the definition of the interaction matrix U on this extended basis set are essential to get results which can be interpreted straightforwardly.
In the second section we will discuss another conceptual extension: A new DMFT self consistent scheme for multilayer systems. Also this step can be understood as the extension of the DMFT basis set. This time, in order to capture features of heterostructures which are beyond the standard DMFT scheme. The main motivation for this implementation are heterostruc-

5 Expanding the basis set

tures similar to those we discussed in chapter 4. However, unlike these heterostructures of chapter, where we encountered the situation of a quasi 2d single layer, we will consider stacks of layers which show non–negligible inter layer hybridization and, as a combined system, display features which are not present in the bulk of the heterostructure constituents.

5.1 LDA+HartreeDMFT

In order to derive the low energy Hamiltonians on a localized basis set from LDA we usually employ methods like NMTO downfolding or Wannier function projections. To capture the essential excitations on the smallest possible basis set these methods are the essential connection between LDA and DMFT as we have seen in the previous chapters. One of the most frequent situations we encounter in transition metal compounds is the incorporation of ligand (e.g. oxygen p) degrees of freedom in *effective* transition metal d–state models. As powerful a tool such downfolding method is, since it gives all the information relevant for the low energy excitations in a very compact form, it turns out to be necessary sometimes to "take a step back" and accept to handle more degrees of freedom on a less entangled basis. One of the main reasons for this is the problematic definition of the interaction matrix U on basis sets which are too compact. The specific problems that arise can be understood intuitively with the help of an example of onsite d– and ligand p– states: If the hybridization of the d– and p– states is small the *effective* d–state Wannier functions will resemble more or less the atomic wave functions which are quite localized. These atomic like wave functions present a good local basis to define the matrix of the local interaction U. Such definition of U is, however, extremely delicate in a situation where the *effective* d–states carry a lot of itinerant p–character. In these cases the Wannier functions are spread widely in space and, hence, they are not a good basis to define the U matrix. Moreover, besides the definition of U, there is a very simple example where a "too effective" model cannot capture well all the important physics:

5.1 LDA+HartreeDMFT

The physics of charge transfer systems are obviously beyond any single site d–states–only model.

5.1.1 Implementation of HartreeDMFT for the nickelates

The best way to introduce our first conceptual extension of the LDA+DMFT approach is probably to discuss an example where we can compare "new" vs. "old" results. In order to do so we will turn back to the nickel–based LaNiO$_3$/LaAlO$_3$– heterostructure of the previous chapter. We recall that the question arose, whether an effective two band model would be sufficient for a thorough treatment of the compound, or if the oxygen p–states could become important when the d–states are split due to U$_{dd}$ interaction effects and the lower Hubbard band is lowered in energy, i.e., shifted towards the p–states. As we stated in the introduction of chapter 4 the final answer for this question is that for physically reasonable values of the interaction parameters the two–band model should be sufficient for a precise description of our nickel–based compounds. In this first part we will proof this conclusion with the HartreeDMFT implementation. In a pedagogical way we can take the following steps:

1. Set up of a suitable dp–model.

2. Solve of the full dp–problem by a *static* mean field Hartree calculation.

3. Finally exchange the Hartree self energies for the most important d–orbitals by a *dynamical* mean field self energy.

5 Expanding the basis set

Let us remark at this point that a combination of DMFT with a Hartree scheme is a very natural step. In fact, it was shown by Müller Hartmann [136] that for the Hubbard model in the DMFT limit of infinite dimensions *"the Hubbard onsite interaction is the only interaction which remains dynamical [...]. All other interactions reduce to their Hartree approximation."* This means, we retain the $d \to \infty$ limit, in which our approximation becomes exact. Furthermore, the p–states are often almost filled and, hence, not strongly correlated. Consequently, this justifies a more approximative treatment of p–related interactions.

Step 1: The dp–model Before directly turning to the model for the nickelates we should consider some general aspects of model Hamiltonains which explicitly take into account ligand p–states. A first obvious feature is an additional site index due to the fact that the Wannier functions of the p–states reside on the ligand atoms. Moreover, if we look at the k–integrated local $\underline{\underline{H}}(R=0)$ Hamiltonian, i.e., the basis in which we write our interaction matrix U, it is no longer diagonal since the dp–basis we choose is not an eigenbasis. Rather we have the following structure:

$$\underline{\underline{H}}_{\text{full}}(R=0) = \int dk \begin{pmatrix} \underline{\underline{H}}_{dd} & \underline{\underline{H}}_{dp} \\ \underline{\underline{H}}_{dp} & \underline{\underline{H}}_{pp} \end{pmatrix} \tag{5.1}$$

However, if the local symmetry allows, the basis can be chosen in a way that at least the $\underline{\underline{H}}_{dd}$ and $\underline{\underline{H}}_{pp}$ blocks are diagonal after k–integration so that the states can be labeled by a good *local* quantum number in the respective subspaces, such as the crystal field labels. This fact is essential for a good definition of the U matrix in the interacting part of the Hamiltonian. This interacting part, again of the multi–orbital Hubbard kind, can capture, besides onsite d– and onsite p–interactions, also possible dp–interactions. Further, we have to realize that, with such definition of the U matrix on a dp–basis the *double counting correction* (DC) of the Hamiltonian derived from LDA becomes more complicated. In the dp–models it does no longer correspond to a simple total energy shift and, hence, cannot be "absorbed" by the chemical poten-

5.1 LDA+HartreeDMFT

tial. We recall the DC in the Anisimov formulation extended to the dp–basis

$$\Delta\varepsilon_{\text{DC}}^{d(p)} = \bar{U}_{dd(pp)}\left(n_{d(p)} - \frac{1}{2}\right) + \bar{U}_{dp}n_{p(d)} \tag{5.2}$$

$n_{d/p}$ being the LDA density of the d– or p– orbital subspace and $\bar{U}_{ml}$ the average interaction value for the respective subspace – see chapter 2. Hence, on the extended basis set the DC generally corresponds to a relative shift of the d– and p–states involved. In the self consistent loop the DC is taken into account at the very beginning since it does not change from iteration to iteration. Let us remark here that, when $n_{d(d)}$ remains unchanged compared to the LDA value the DC term cancels the effect of U_{pd} exactly.

After these general considerations let us now turn to the specific case of the dp–model for the nickelate heterostructure. The dp–model was obtained by NMTO downfolding and consists, besides the two e_g–orbitals, of three additional bands reflecting the p–degrees of freedom ($p_x; p_y; p_z^{b\cdot}$ where $p_z^{b\cdot}$–denotes a bonding linear combination of the two apical ligands). In order to write down also an analytical expression, the hopping was truncated in real space to nearest and next nearest neighbor d–p hoppings taking also p–p hopping into account. In the upper panel of Fig. 5.1 we show a cartoon of our model together with the corresponding non–interacting band structure and density of states (d–states in black/light gray; sum of the three p–states: dark gray). From the band structure and DOS plot in the lower panels we see that there is significant hybridization between the p– and d–states. As can be understood intuitively from our sketch the $3z^2 - r^2$ states (light gray) hybridize strongest with the bonding $p_z^{b\cdot}$ orbital corresponding to the peaks around 1.8 eV and -3.4 eV, whereas the $x^2 - y^2$ states (black) in our model hybridize *only* with the planar ligands corresponding to the peaks at -0.8 eV and -4.0 eV: the overlap of the $p_z^{b\cdot}$ and the $x^2 - y^2$ states is zero (no black peak at -3.4 eV). Further, we observe a hybridization between the two e_g states similarly to the two band model; this time, however, the e_g–hybridization is explicitly mediated via d–p–d hopping processes. The Hamiltonian for this model and its hopping amplitudes (up to next nearest neighbor) are summarized in

5 Expanding the basis set

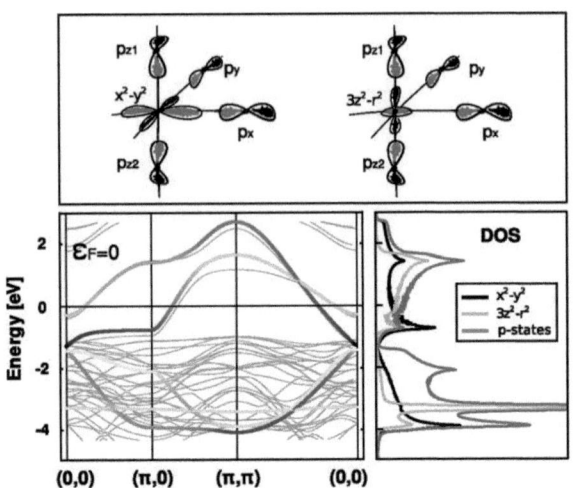

Fig. 5.1: Upper panel: Cartoon of the five–band model for the nickelate heterostructures including two e_g orbitals and three p orbitals explicitly. Lower panels: Tight binding band structure (right hand side) with truncated hopping (compare tables 5.1 and 5.2). The shading encodes the orbital character for $x^2 - y^2$ (black), $3z^2 - r^2$ (light gray), and p–states (dark gray). On the left hand side we plot the corresponding partial DOS.

Tab. 5.1 and 5.2.

The next step is to obtain a self consistent but *static* mean field solution of the problem.

Step 2: Hartree solution of the full problem The solution of the dp–problem by means of "LDA+Hartree" is, in the end, almost identical with a (ferromagnetic) LDA+U calculation with the difference that we perform the calculation already on the downfolded/projected dp–basis. In order to perform a self consistent Hartree solution we recall that we could employ the same self consistent way as the DMFT depicted in Fig. 2.6 in chapter 2. However, instead of the DMFT, we use here the very simple Hartree approach to solve the

5.1 LDA+HartreeDMFT

	$\|x^2-y^2\rangle$	$\|3z^2-y^2\rangle$	$\|p_x\rangle$	$\|p_y\rangle$	$\|p_z\rangle$
	ε_1	0	$-\mathrm{i}v_{1x}$	$\mathrm{i}v_{1y}$	$2\sqrt{2}t_{15}^2 v$
	0	ε_2	$\mathrm{i}v_{2x}$	$\mathrm{i}v_{2y}$	$-\sqrt{2}(t_{25}^1+2t_{25}^2)$
	$\mathrm{i}v_{1x}$	$-\mathrm{i}v_{2x}$	$\varepsilon_3 - 4t_{33}^3\cos(k_x)\cos(k_y)$	$4t_{34}^1\sin(\tfrac{1}{2}k_x)\sin(\tfrac{1}{2}k_y)$	$-\mathrm{i}\sqrt{2}v_{3x}$
	$-\mathrm{i}v_{1y}$	$-\mathrm{i}v_{2y}$	$4t_{34}^1\sin(\tfrac{1}{2}k_x)\sin(\tfrac{1}{2}k_y)$	$\varepsilon_3 - 4t_{33}^3\cos(k_x)\cos(k_y)$	$-\sqrt{\mathrm{i}}2v_{3y}$
	$2\sqrt{2}t_{15}^2 v$	$-\sqrt{2}(t_{25}^1+2t_{25}^2)$	$-\mathrm{i}\sqrt{2}v_{3x}$	$-\sqrt{\mathrm{i}}2v_{3y}$	$\varepsilon_5 + 2t_{55}^2 u$

Tab. 5.1: Tight binding Hamiltonian for the five band model sketched in Fig. 5.1 – (hopping is truncated after next nearest neighbors and basis functions are chosen real). From Xiaoping Yang & O.K. Andersen (MPI Stuttgart)

$$u = \cos(k_x) + \cos(k_y) \qquad v = \cos(k_x) - \cos(k_y)$$

$$v_{1x} = 2t_{13}^1 \sin(\tfrac{1}{2}k_x) + 4t_{13}^2 \sin(\tfrac{1}{2}k_x)\cos(k_y) \qquad v_{2x} = 2t_{23}^1 \sin(\tfrac{1}{2}k_x)$$

$$v_{3x} = 2t_{35}^1 \sin(\tfrac{1}{2}k_x) \qquad v_{1y} = 2t_{13}^1 \sin(\tfrac{1}{2}k_y) + 4t_{13}^2 \sin(\tfrac{1}{2}k_y)\cos(k_x)$$

$$v_{2y} = 2t_{23}^1 \sin(\tfrac{1}{2}k_y) \qquad v_{3y} = 2t_{35}^1 \sin(\tfrac{1}{2}k_y)$$

$t_{13}^1 = -1.0478$	$t_{13}^2 = 0.0667$	$t_{15}^2 = -0.0589$	$t_{23}^1 = -0.6152$
$t_{25}^1 = -1.1840$	$t_{25}^2 = 0.0526$	$t_{33}^3 = -0.0841$	$t_{34}^1 = -0.3309$
$t_{35}^1 = -0.3381$	$t_{55}^2 = -0.0806$		

Tab. 5.2: Parameters and hoppings for Hamiltonian 5.1

5.1 LDA+HartreeDMFT

AIM in the larger basis set. The *static* mean field self energy consists only of the Hartree–balloon diagram and is, hence, only a function of the densities of the interacting orbitals[1]:

$$\Sigma^H_{mm} = \sum_l U_{lm} \cdot <n_l> \qquad (5.3)$$

or diagrammatically

$$\Sigma^H = \text{\raisebox{-0.5ex}{\includegraphics[scale=0.5]{balloon.pdf}}} \qquad (5.4)$$

This method is computationally not very expensive and sometimes can yield, a good first glimpse on the relevant excitations. However, because the opening of a gap in such static mean field theory always needs breaking of symmetry the tendencies towards spin or orbital ordering are strongly overestimated. Therefore the Hartree results should be handled cautiously.

For the definition of the U matrix of our nickelate dp–model we choose parameters describing a strongly correlated situation and estimate the values to be in a physically reasonable range: $U_{dd} = 8.0\text{eV}$, $U_{pp} = 4.0\text{eV}$ and we consider different values of U_{pd}. Let us briefly remark that the values for the U matrix in a dp–model have to be chosen larger in comparison to effective d–only models, since screening effects, i.e., the rearrangement of p– and d–states with respect to one another are now taken into account explicitly. Of the above mentioned parameters the U_{pd} is the least obvious parameter to estimate, since it depends in a very involved way on the spread of the d– and p–wave functions and their respective overlap. One could speculate, that the less localized p– wave functions "swallow" the d– wave functions (compare the Wannier function plots in Fig. 5.6) so that a reasonable choice may be $U_{pd} \lesssim U_{pp}$. Further, we stress that the effect of U_{pd} corresponds directly to a

---

[1] In fact there is no need to employ the DMFT–style mapping to the AIM at all since the Hartree decoupling can directly be carried out on the LDA–constructed Hamiltonian. Nonetheless, since we want to combine the Hartree and the DMFT approach later, the formulation of the Hartree scheme in terms of the self consistent solution of the AIM turns out to be the most useful one.

5 Expanding the basis set

"rigid" potential shift of d– and p–states in the limit of small hybridization. Within a full self consistent calculation, however, it has more subtle effects since the value of the self energy originating from U_{pd} depends also on the correlation induced charge rearrangement. Hence, we understand that these effects are *not* included in the LDA results and should be taken into account explicitly. In conclusion we state, that the part of the self energy originating from U_{pd} corresponds to a *self consistently* calculated d–p splitting.

In order to keep also the interaction parameterization simple for the present case we do not take into account a dp–spin coupling J_{dp} but only an onsite Hund's coupling J_{dd}. After the self consistent calculation of the Hartree self energy we obtain the local spectral function by calculating for each diagonal element

$$A(\omega)_{m,m} = \Im\left[G^{\text{loc.}}_{m,m}(\omega)\right] = \Im\left[\frac{1}{V_{\text{BZ}}}\int_{\text{BZ}} d^3k \frac{1}{\omega + \mu - \varepsilon(\mathbf{k}) - \Sigma^{\text{H}} + i\delta \hat{\mathbb{1}}}\right]_{m,m} \quad (5.5)$$

where we directly employed the analytical continuation explicitly substituting $i\omega_\nu \to \omega + i\delta$ with small δ.

In Fig. 5.2 we show the resulting local spectral functions ($U_{dd} = 8.0$eV, $U_{pp} = 4.0$eV) for four different values of the U_{pd}. The Fermi energy $\varepsilon_F = 0$ has been set to zero (dashed line) in these plots. We show partial spectral functions for which the color indicates the respective character: For the $x^2 - y^2$ and the $3z^2 - r^2$ spectra we choose the same colors, i.e. black and light gray, as in chapter 4, whereas the three p–spectra have been summed up and are plotted in a dark gray tone. As the spectra show, the solutions for our set of parameters are all insulating, i.e., have a gap at the Fermi energy. In the static Hartree mean field spectra this gap is an *artificial* spin gap. A symmetry breaking with respect to the spin is, besides orbital ordering, the only way to obtain a non trivial[2] convergence in the self consistent equations.

Leaving the problem of the "fake ordering" aside, we start our discussion with the spectrum for the smallest value, i.e. zero, for U_{pd} in the top left panel. We observe that the d–spectrum is gapped and forms "Hubbard bands" sepa-

---
[2] i.e. different from the LDA solution

5.1 LDA+HartreeDMFT

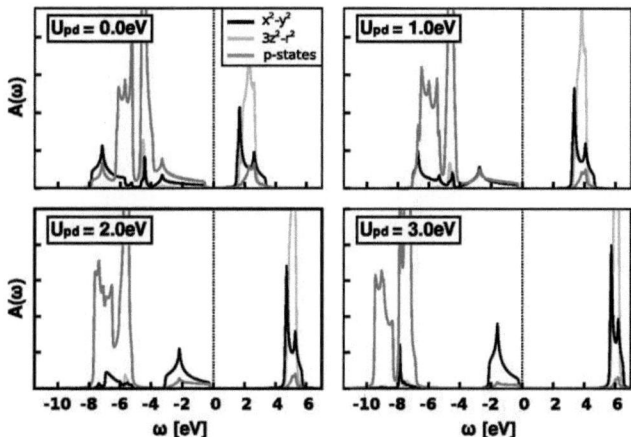

Fig. 5.2: Static mean field Hartree spectral functions for the five–band model 5.1 calculated for $U_{dd} = 8.0$eV, $U_{pp} = 4.0$eV and four different values of U_{pd}. The Fermi energy $\varepsilon_F = 0$ has been set to zero (dashed line) in these plots and the coloring encodes the orbital character. In order to get non–trivial convergence the symmetry had to be broken – all solutions are ferromagnetic.

5 Expanding the basis set

rated roughly by an energy of the order of U_{dd} at around -7 eV and 2 eV. The lower Hubbard band is strongly hybridizing with the p–states which reside, very broad, in the region of -8 eV to -1 eV. Also realize, that the d–spectrum below ε_F mainly consists of $x^2 - y^2$ character (black), while the $3z^2 - r^2$ contribution (light gray) is almost entirely located at around 2 eV above ε_F – this observation is in perfect agreement with the LDA+DMFT results of the previous chapter, although the present dp–model, due to its truncated hopping, is obviously less accurate in reproducing specific details of the band structure. However, the pure Hartree treatment shall be considered just as a first approximation. On the one hand, as mentioned before, it can describe spectral gaps only by ordering, and on the other hand the "static" nature of this approximation does not allow for a correct description of excited states leading to systematic errors in the spectral function even in broken symmetry phases with large U [181]. Nonetheless, in the light of the discussions of chapter 4 we can interpret the Hartree result as an indication that the LDA+DMFT results are quite robust. Turning back to the spectrum for $U_{pd} = 0.0$ eV we find that the actual gap in the total spectrum is a charge transfer gap of the order of 2 eV. Yet, we should include a finite value of U_{pd}. In the remaining three panels we show the evolution of the spectra upon increasing U_{pd}. The main effect of the U_{pd} that we can observe is a suppression of the hybridization. Already at a value of $U_{pd} = 2.0\text{eV} = 1/2U_{pp} = 1/4U_{dd}$ and even more so for $U_{pd} = 3.0$ eV we find that the d– and p–states become almost completely separated and the gap is a generic d–states Mott gap. This result is the first indication that the two band model should be sufficient in order to treat the system in the vicinity of ε_F for reasonable interaction parameters. In order to find further support for this claim we now turn to step 3 of our pedagogical discussion: The actual HartreeDMFT implementation.

Step3: HartreeDMFT The third and last step is very much straightforward. In this step we are going to replace the crude Hartree approximation by the much more accurate DMFT scheme for the *most important correlations* in the system, i.e., the onsite d–correlations. In other words we treat *all* correlations U_{dd}, U_{pp}, and U_{pd} but assume that the smaller effects of U_{pp} and U_{pd} can be captured only by the static Hartree approach, while the U_{dd} is treated with the DMFT self consistent solution of AIM problems. With this separation we achieve a great improvement of our results: We incorporate completely the advantages of the DMFT albeit keeping the p–degrees of freedom. This means we can abolish the artificial aspect of spin or orbital orderings and, moreover, we can now describe the entire range from weak to strong correlations. With the DMFT we obtain a much better description of the excited states and, further, the extended basis set now allows us not only to capture the physics of Mott–Hubbard–, but also charge–transfer insulators and the concomitant d–p interplay.

The HartreeDMFT implementation can also be seen as an embedding of a DMFT calculation on a larger basis allowing for a coupling between the strongly correlated (d–states) and less correlated degrees of freedom (p–states). Let us stress that this kind of embedding is also the first step towards a fully self consistent LDA+DMFT scheme in which a DMFT self energy is calculated *within* each LDA iteration.

In order to fully comprehend the implementation of the algorithm, we will rewrite the equations of the self consistency as an extension of the equations in chapter 2. Before the self consistency loop we have to perform an initialization which consists of performing the double counting correction (5.2) and choosing a starting self energy. Let us remark here that it is necessary to separate the self energy into two components: The full self energy is the sum of a part Σ^H associated to the U_{pp} and U_{pd}, which will be calculated with the *static* Hartree scheme, and a part Σ^{DMFT} associated to the U_{dd}, which is calculated with DMFT. Since we will employ the Hirsch Fye QMC as the impurity solver for the calculation of Σ^{DMFT} we have an additional correction to perform: The Hubbard–Stratonovic shift (see Info box below)

After the initialization we write the self consistent loop:

1. The first step is, also for the HartreeDMFT, the calculation of the k–integrated Green function – this time on the *full* d–p basis set:

5 Expanding the basis set

INFO: The Hubbard–Stratonovic shift in Hirsch Fye QMC

The Hirsch Fye QMC algorithm decouples the two–particle operator of the Coulomb interaction by a Hubbard–Stratonovic transformation (HST). For the simplification of the decoupling we have to pay with an additional summation over an auxiliary field of "spins" which is done by Monte Carlo sampling. The crucial point is that for the HST we decouple a term which reads for the one band case:

$$\frac{U}{2} \sum_{\sigma,\sigma',\sigma \neq \sigma'} (n_\sigma + n_{\sigma'})^2 = \frac{U}{2} \sum_{\sigma,\sigma',\sigma \neq \sigma'} \left(-2 n_\sigma n_{\sigma'} + n_\sigma^2 + n_{\sigma'}^2 \right) \tag{5.6}$$

where $n_\sigma^2 = n_\sigma$. The first summand corresponds to the interaction part of the Hubbard Hamiltonian which means that we have to compensate for an additional term:

$$\frac{U}{2} \sum_{\sigma,\sigma',\sigma \neq \sigma'} (n_\sigma + n_{\sigma'}) = \frac{U}{2} \cdot N \tag{5.7}$$

N being the total number of interacting electrons. The associated energy change can then be calculated as the derivative of (5.7) with respect to the number of electrons N, yielding

$$\Delta \varepsilon_{\text{1band}}^{\text{HS}} = \frac{\partial}{\partial N} \frac{U}{2} \cdot N = \frac{U}{2} \tag{5.8}$$

In a general multiband case with orbital dependent U the expression has to be extended correspondingly:

$$\Delta \varepsilon_{m,\sigma}^{\text{HS}} = \frac{1}{2} \sum_{m',\sigma'}{}' U_{m,m'}^{\sigma,\sigma'} \tag{5.9}$$

where the prime above the sum denotes $m\sigma \neq m'\sigma'$ and $U_{m,m'}^{\sigma,\sigma'}$ is the interaction parameter of the m,σ electrons with m',σ' electrons. As mentioned before $\Delta\varepsilon_{HS}$ can be neglected in a d–states only DMFT when the interaction parameter U is the same for all orbitals, since it can be absorbed by the chemical potential. However, as soon as we treat the d–d interaction differently compared to d–p or p–p, like in the HartreeDMFT we must take this shift (of the d–potential) into account explicitly.

$$\underline{\underline{G}}_{\text{full}}^{\text{loc.}}(\omega) = \frac{1}{V_{\text{BZ}}} \int_{\text{BZ}} d^3k \left[(\omega+\mu)\underline{\underline{\mathbb{1}}} - \underline{\underline{\varepsilon}}(\mathbf{k}) - \underline{\underline{\Sigma}}_{dp}(\omega)\right]^{-1} \quad (5.10)$$

where the double underlines denote matrices on the dp–basis.

2. Next, we separate the local Green function for the d–subspace:

$$\underline{\underline{G}}_{\text{d}}^{\text{loc.}}(\omega) = \mathcal{P}_{\text{d}} \, \underline{\underline{G}}_{\text{full}}^{\text{loc.}}(\omega) \quad (5.11)$$

where $\mathcal{P}_{\text{d}}$ is the Projector on the d–subspace. We stress here, that due to the d–p hybridization, encoded in the Hamitonian $\underline{\underline{\varepsilon}}(\mathbf{k})$ and the inversion of Eq. 5.10 the information about the p–ligands is not lost but captured by $\underline{\underline{G}}_{\text{d}}^{\text{loc.}}(\omega)$.

3. Now, in complete analogy to d–states only DMFT, we calculate the Weiss field for the impurity model (only on the d–subspace):

$$\left[\underline{\underline{\mathcal{G}}}^0(i\omega_\nu)\right]^{-1} = \left[\underline{\underline{G}}_{\text{d}}(\omega)\right]^{-1} + \underline{\underline{\Sigma}}_{\text{d}}^{\text{DMFT}}(\omega) \quad (5.12)$$

where $\underline{\underline{\Sigma}}_{\text{d}}^{\text{DMFT}}(\omega)$ denotes *only* the DMFT part of the full self energy $\underline{\underline{\Sigma}}_{dp}(\omega)$. This means that the static part of the self energy Σ^{H}, which is not explicitly taken into account, enters the AIM nonetheless. Its information, the

5 Expanding the basis set

respective Hartree potentials for each orbital, is contained in the Weiss field.

4. With $\underline{\underline{\mathcal{G}}}^{\sigma 0}_{lm}(i\omega_\nu)$ we, exactly as before, calculate a new $\underline{\underline{G}}^{loc.}_{d}$ and with the inverse of Eq. (5.12) we obtain a new $\underline{\underline{\Sigma}}^{DMFT}_{d}(\omega)$.

5. This leaves us with the calculation of the *static* Σ^H for which we simply have to calculate the orbital dependent density and apply Eq. (5.3).

6. Now we have to assemble the new $\underline{\underline{\Sigma}}^{new}_{dp}(\omega) = \underline{\underline{\Sigma}}^{DMFT}_{d}(\omega) + \underline{\underline{\Sigma}}^{H} - \underline{\underline{\varepsilon}}_{HS}$ where we also take into account the HS shift (see **Info**). Finally we close the self consistent loop by comparing $\underline{\underline{\Sigma}}^{new}_{dp}(\omega)$ and $\underline{\underline{\Sigma}}^{old}_{dp}(\omega)$ and iterating until convergency is reached.

In Fig. 5.3 we show a compact sketch of our HartreeDMFT scheme. We start on the top left with the choice of a starting self energy which enters the self consistent loop to obtain a converged solution after a sufficient number of iterations.

Let us now discuss the results of the HartreeDMFT for the nickelate model. In Fig. 5.4 we show the spectral functions for the dp–model. The interaction parameters for the HartreeDMFT were chosen slightly larger than in the Hartree calculation due to the above mentioned differences of the static and dynamical mean field theories [181]. In this case we choose the parameters $U_{dd} = 10$ eV, $U_{pp} = 5$ eV, and again four different values for U_{pd}. The most striking observation is that we now obtain a metal to insulator transition (MIT) upon increasing the value of U_{pd}. Moreover, our results are fully paramagnetic without ordering. Such analysis of the MIT from a correlated metal to an insulating state is impossible by means of the static Hartree calculation. The HartreeDMFT spectral function for $U_{pd} = 0$ eV shows strong dp–hybridization

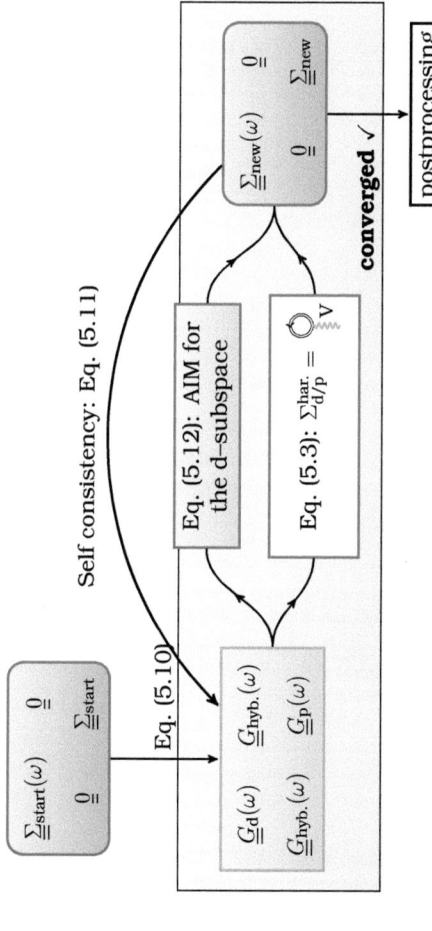

Fig. 5.3: Scheme of the HartreeDMFT implementation. The differences to standard DMFT are the projection before the calculation of the impurity problem Weiss field for the d–subspace and the Hartree treatment of less correlated degrees of freedom (white box)

5 Expanding the basis set

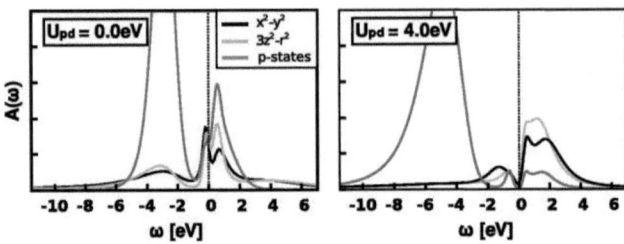

Fig. 5.4: LDA+HartreeDMFT spectral functions for the five band model calculated for $U_{dd} = 10$ eV, $U_{pp} = 5$ eV, and $U_{pd} = 0$ eV (left hand side) or $U_{pd} = U_{pp}$ (right hand side). The interaction parameters for the HartreeDMFT were chosen slightly larger than in the Hartree calculation due to the differences of the static and dynamical mean field theories [181].

at ε_F and as a result a not so strongly correlated, metallic, spectrum for the d–states with coherent quasi particle excitations. However, if we restrict the charge transfer between d– and p–states by means of a (reasonably chosen) U_{pd}, assuming that it is not *for free* to move the electrons around, we obtain very strongly correlated and even insulating solutions. The features of the insulating solution is in qualitative agreement with the Hartree results shown in Fig. 5.2 (apart from the artificial ordering in Hartree, of course), and thus, the results for the two–band model of the previous chapter. The $3z^2 - r^2$ band is pushed above the Fermi energy due to the enhancement of the "effective crystal field splitting" and the conclusions discussed in chapter 4 remain valid.

In this section we presented the implementation of the so called HartreeDMFT scheme using the example of a dp–model for the nickelates. We have learned how a full self consistent loop is set up and how certain interactions can be treated separately with more (DMFT) or less (Hartree) expensive calculations depending on their importance for the actual low energy excitations of the system.

5.1.2 The cuprate "(extended) Emery" model: a revision

Let us now turn to a system, where we expect the dp–model can really yield different results from an effective d–model: the high T_C cuprates.
A popular tight–binding Hamiltonian for the high T_C cuprates is a 3–band model suggested by Emery in 1987 [54]. Let us review the history of this model and also point out the connection to the *planar/axial*-orbital model for the cuprates which we discussed in the first section of chapter 4. The Emery model consists of one planar Cu $x^2 - y^2$ band, two oxygen p_x and p_y bands and takes into account a d–p hopping. It is thus the minimal model in order to describe the charge transfer insulating state and, moreover, the physics of a Zhang Rice singlet [222]. These features are obviously beyond a description of the cuprates within an effective single band model. However, Andersen *et al.* [5] concluded from downfolding *ab initio* LDA bandstructure that the original model as described in [54] should be extended. The extensions that are proposed in [5] originate basically from the inclusion of the *axial* degree of freedom which was shown to be the material–dependent quantity by Pavarini *et al.* [154]. Specifically, the work of Andersen *et al.* starts by constructing an eight band Hamiltonian from which all high energy degrees of freedom have been integrated out. The eight orbitals of this Hamiltonian are separated into a 4×4 σ–bonding block of Cu $3d_{x^2-y^2}$, O_1 $2p_x$, O_2 $2p_y$, as well as Cu 4s (with some Cu $3d_{3z^2-r^2}$ character), and another 4×4 π–bonding block of Cu $3d_{xz}$, Cu $3d_{yz}$, O_1 $2p_z$, and O_2 $2p_z$. The generic Hamiltonian for the CuO$_2$ planes is the 4×4 σ–block, since it contains the conduction band. Further, due to symmetry reasons the σ– and the π–block do not hybridize in the limit of *flat planes* [5].
Within the four band σ–Hamiltonian the Cu 4s plays the special role of the *axial* state reflecting the material–dependence as we know from our previous studies (see chapter 4). In the two band *planar/axial* model the degrees of freedom of the σ–Oxygen states were folded into the *planar* $x^2 - y^2$ conduction band. The three band Emery–like model is, instead, obtained by folding the Cu 4s *axial* degrees of freedom down to the oxygen bands, resulting in an additional p–p hopping and a renormalization of the onsite p–energy.

5 Expanding the basis set

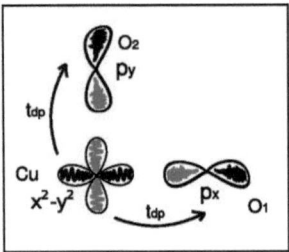

Fig. 5.5: Cartoon of the "historic" Emery model (1987 [54]) including only d–p hopping

This means that the material dependence we described in chapter 4 is now encoded in the tails of the oxygen bands, i.e., the renormalized hopping parameters and specifically the t_{pp}. The resulting Hamiltonian can be written analytically and we give its elements in Tab. 5.3.

Motivation We performed HartreeDMFT calculations for the undoped La_2CuO_4 (LSCO) compound[3]. The analysis was mainly motivated by inconsistencies of the results and the parameter choices of other recent studies.

Let us start by reporting the NMTO downfolding results and the corresponding hopping parameters of Hamiltonian in Tab. 5.3. In Figs. 5.6 & 5.7 we show the NMTO bands from Saha–Dasgupta *et al.* [179] plotted on top of the LDA bands (thin lines) for N=0 (Fig. 5.6) and N=1 (Fig. 5.7) as well as the corresponding hopping parameters. The color codes the orbital character: 3d $x^2 - y^2$ is black, whereas the p–states are summarized and plotted in gray – color mixing corresponds to orbital hybridization. The difference of the N=0 and N=1 both models is the energy range in which they reproduce the cuprate LDA bandstructure (thin lines): While the N=0 model only has one fixed energy, namely the Fermi energy ε_F, the N=1 model was fixed to ε_F and to the energy of the bottom of the oxygen p–bands at around -8 eV. By having to span a wider energy range, the N=1 orbitals are somewhat less localized, and consequently have longer–ranged hoppings, than the N=0 orbitals. A thorough discussion on the relations and trends of the hopping parameters

---
[3]Recall that LSCO is a strongly correlated insulator for which the LDA yields a metallic ground state

can be found in [103] and [179]. In our study we will consider first the N=0 and then the N=1 model in order to address an issue of central interest:
The most problematic parameter in recent studies [49, 213] was the choice of the d–p splitting $\varepsilon_d - \varepsilon_p = \Delta_{dp}$. While the NMTO downfolding yields a value of $\Delta_{dp} = 0.45$ eV (N=0) or $\Delta_{dp} = 0.96$ eV (N=1) it turns out that the many–body treatments which include correlation effects fail to reproduce the insulating behavior of the undoped LSCO so that, in order to fix this problem, the Δ_{dp} was increased "by hand" to values of the order of $\Delta_{dp} \approx 3$ eV [103, 213] or it was chosen as a variable parameter [49]. Further, Kent et al. [103] pointed out that previous justifications of such enhancement of Δ_{dp} by means of constrained LDA calculations for La_2CuO_4 are problematic due to a misleading assumption of the electron count (for details see the discussion in [103] page 3 left column). Our HartreeDMFT results for both (N=0 and N=1)MTO models can be summarized in two conclusions:

1. The d–p interaction U_{pd} which leads to a *self consistently determined* level shift can drive the system insulating within the HartreeDMFT approach and open the charge transfer gap (including the Zhang Rice singlet states).

2. The critical values of the interaction parameters U_{ml} for the MIT are quite large for the *original* NMTO set ([103] and [179]) of parameters. With this observation we can actually speculate that the insulating antiferromagnetic ground state of La_2CuO_4 is, in fact, somewhat less strongly correlated and non–local correlations, which are beyond the single site DMFT, have to be taken into account for a thorough description.

La_2CuO_4: HartreeDMFT results Let us start with the results for the (N=0)MTO model (see bands in Fig. 5.6 left hand side). In order to illustrate the model we show the Wannier functions related to the (N=0)model in Fig. 5.6 upper panel: Centered on the Cu sites we find the 3d Cu $x^2 - y^2$ orbital and the

5 Expanding the basis set

$$\hat{H}_{\text{pd}}^{\text{LDA}} = \sum_{\mathbf{k}lm\sigma} h_{lm}^{\text{LDA}}(\mathbf{k}) c_{l\sigma}^{\dagger}(\mathbf{k}) c_{m\sigma}(\mathbf{k})$$

$$\underline{\underline{h}}^{\text{LDA}}(\mathbf{k}) \begin{pmatrix} h_d(\mathbf{k}) & h_{d,p1}(\mathbf{k}) & h_{d,p2}(\mathbf{k}) \\ h_{p1,d}(\mathbf{k}) & h_{p1}(\mathbf{k}) & h_{p1,p2}(\mathbf{k}) \\ h_{p2,d}(\mathbf{k}) & h_{p2,p1}(\mathbf{k}) & h_{p2}(\mathbf{k}) \end{pmatrix}$$

$$h_d(\mathbf{k}) = \varepsilon_d + 2t_{dd}(\cos(k_x) + \cos(k_y)) + 4t'_{dd}\cos(k_x)\cos(k_y)$$

$$h_{p1}(\mathbf{k}) = \varepsilon_p + 2.0(t'_{pp}\cos(k_x) + t''_{pp}\cos(k_y) + 2t''''_{pp}\cos(k_x)\cos(k_y))$$

$$h_{p2}(\mathbf{k}) = \varepsilon_p + 2.0(t'_{pp}\cos(k_y) + t''_{pp}\cos(k_x) + 2t''''_{pp}\cos(k_y)\cos(k_x))$$

$$h_{d,p1}(\mathbf{k}) = 2((t_{pd} + 2t'_{pd}\cos(k_y))\sin(k_x/2) + (t''_{pd} + 2t'''_{pd}\cos(k_y))\sin(3k_x/2))$$

$$h_{d,p2}(\mathbf{k}) = -2.0((t_{pd} + 2t'_{pd}\cos(k_x))\sin(k_y/2) + (t''_{pd} + 2t'''_{pd}\cos(k_x))\sin(3k_y/2))$$

$$h_{p1,p2}(\mathbf{k}) = -4.0(t_{pp}\sin(k_x/2)\sin(k_y/2) + t''''_{pp}(\sin(3k_x/2)\sin(k_y/2)$$
$$+ \sin(3k_y/2)\sin(k_x/2)))$$

Tab. 5.3: Extended Emery model [5] including p–p hopping mediated by the *material dependent* axial degree of freedom.

5.1 LDA+HartreeDMFT

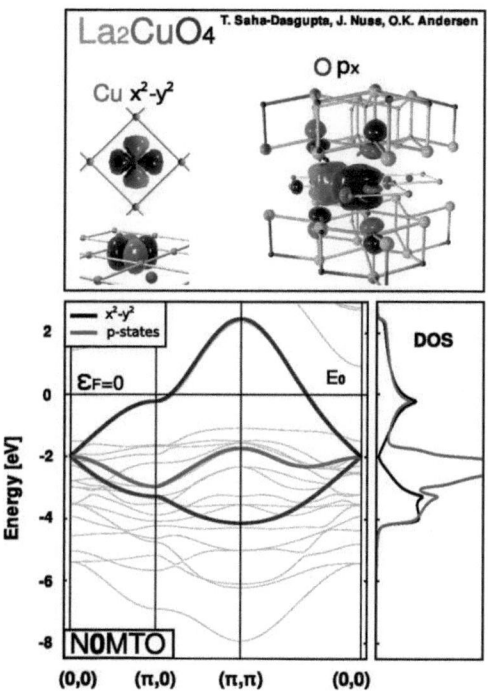

$\varepsilon_d - \varepsilon_p = 0.43\text{eV}$	$t_{dd} = -0.10\text{eV}$	$t_{pd} = 0.96\text{eV}$	$t'_{pd} = -0.1\text{eV}$
$t_{pp} = 0.15\text{eV}$	$t'_{pp} = -0.24\text{eV}$	$t''_{pp} = 0.02\text{eV}$	$t'''_{pp} = 0.11\text{eV}$

Fig. 5.6: (N=0)MTO model for La_2CuO_4. Top panel: (N=0)MTO orbitals for the Cu 3d $x^2 - y^2$ states (left hand side) and the O 2p x, y states (right hand side). The oxygen p–states "swallow" the copper sites calling for the consideration of a dp interaction U_{pd}. Lower panel: (N=0)MTO band structure (left hand side) on top of the LMTO bands (thin lines) and the corresponding density of states (right hand side). The shading encodes the band character. Bottom table: all considered hoppings taken from Kent *et al.* [103]

5 Expanding the basis set

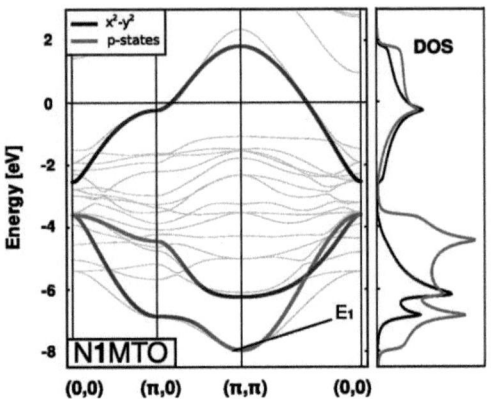

Fig. 5.7: (N=1)MTO model for La$_2$CuO$_4$. Top panel: (N=1)MTO band structure (left hand side) on top of the LMTO bands (thin lines) and the corresponding density of states (right hand side). The shading encodes the band character. Bottom table: all considered hoppings taken from Kent *et al.* [103]

$\varepsilon_d - \varepsilon_p = 0.95$eV $\quad t_{dd} = 0.15$eV $\quad t_{pd} = 1.48$eV $\quad t'_{pd} = 0.08$eV
$t_{pp} = 0.91$eV $\quad t'_{pp} = 0.03$eV $\quad t''_{pp} = 0.15$eV $\quad t'''_{pp} = 0.03$eV

162

5.1 LDA+HartreeDMFT

O 2p–functions centered on the (violet) oxygen sites. From this picture we intuitively understand that a rearrangement of the d–charge should couple back to the oxygen p–states without neglecting their respective repulsion: The oxygen 2p–function "swallows" the d–site almost entirely; p and d states are hence close by and their mutual repulsion, i.e. U_{pd}, becomes relevant. In the lower panel we show the Wannier functions for $HgBa_2CuO_4$ which present an even stronger overlap. As a side remark we recall that the difference between La_2CuO_4 and $HgBa_2CuO_4$ is the position of the *axial* Cu 4s band, which in the three band model is reflected in the tails of the p–states which we can observe in Fig. 5.6.

We performed the HartreeDMFT calculations in the same way as described in the previous section. For the interaction parameters we choose $U_{dd} = 10$ eV, $U_{pp} = 5$ eV, and take different values of U_{pd} ranging from $U_{pd} = 0$ eV to $U_{pd} = U_{pp} = 5$ eV. The resulting spectral functions are shown in Fig. 5.8. As in the bandstructure plots, the color of the spectral function codes the orbital character. Let us start from the spectrum for $U_{pd} = 0$ eV which is plotted in the top left panel. Although the parameters $U_{dd} = 10$ eV and $U_{pp} = 5$ eV are by no means small compared to the bandwidth, we observe a rather uncorrelated spectral function which resembles the non–interacting DOS (Fig. 5.6). The main reason for this is that, due to the d–p hybridization, the filling of each band and, above all of the d–band, is far from an integer value. Upon increasing the value for U_{pd} this hybridization decreases, as it can be seen in the spectra, since the charge transfer from d–states to p–states is now connected with a potential shift of the respective states of the order of U_{pd}. Eventually, for values $U_{pd} \approx U_{pp}$, we observe a rather sudden metal to insulator transition between $U_{pd} = 3$ eV and $U_{pd} = 4$ eV. The last two spectra in the bottom panels of Fig. 5.8 show a gap between an d–states "upper Hubbard band" with some p–hybridization above the Fermi energy ε_F and a mixed d–p peak around -1 eV (for $U_{pd} = 4$ eV) or -2 eV (for $U_{pd} = 4$ eV). This mixed peak can, in fact, be associated with the Zhang Rice states (Cu d–hole, O p–hole pair[4]) and have been also observed also in the previous studies [49, 213] Δ_{dp}

---

[4]However, in the paramagnetic calculations these states are not a singlet.

5 Expanding the basis set

was taken as a "free" parameter. The "lower Hubbard band" is quite broad and centered around ~ -7 eV, whereas most of the p–spectral weight is located in a large peak around -4 eV ($U_{pd} = 4$ eV) and -5 eV ($U_{pd} = 5$ eV).
In summary, the HartreeDMFT yields an insulating state for the original (N=0)MTO parameters *without* the artificial enhancement of the d–p splitting. Instead, we assumed a finite value of U_{pd} which, in a self consistent way, leads to a suppression of d–p hybridization driving the metal to insulator transition. Yet, we should remark two issues: First of all the transition occurs at rather large values for the interaction parameters and, secondly, we have to realize that the (N=0)MTO was designed to reproduce the cuprate bands only around the Fermi energy [103]. Hence, it is questionable if the (N=0)MTO really yields a good basis for a study of excitations on an energy scale of some eV above and below the Fermi energy such as the d–p interplay. Therefore, let us turn to the HartreeDMFT results of the (N=1)MTO model.

We recall that the N=1 model has been fixed to the Fermi energy ε_F and, moreover, to an energy at the bottom of the oxygen p–bands at around -8 eV (Fig. 5.7). As mentioned above, the N=1 orbitals are less localized, and consequently have somewhat longer–ranged hoppings, than the N=0 orbitals in order to reproduce the band structure on a larger energy range. Hence, the values for the interaction parameters should be correspondingly reduced as a consequence of the less localized character of the orbitals. Yet, it turned out, that for the (N=1)MTO we obtain much more *metallic solutions*. The system remains metallic even for the same interaction parameters as we used for the N=0 model. In Fig. 5.9 we show the HartreeDMFT spectral functions for the N=1 model: In the left panel, we plotted the spectrum for the parameter set $U_{dd} = 10$ eV, $U_{pp} = 5$ eV, and $U_{pd} = 5$ eV, i.e., the parameter set where the N=0 model already showed a gap of ≈ 1 eV. In order to obtain an insulating solution (see right panel of Fig. 5.9), we had to increase the interaction parameters to values of $U_{dd} = 13$ eV, $U_{pp} = 7$ eV, and $U_{pd} = 7$ eV, which are physically no longer justifiable. Moreover, the insulating spectrum no longer displays the physics of a charge transfer insulator, since the Cu d–"lower Hubbard band" is *above* the O p–spectrum! Conclusively we state that the results for the N=1 model within the HartreeDMFT approach are not in agree-

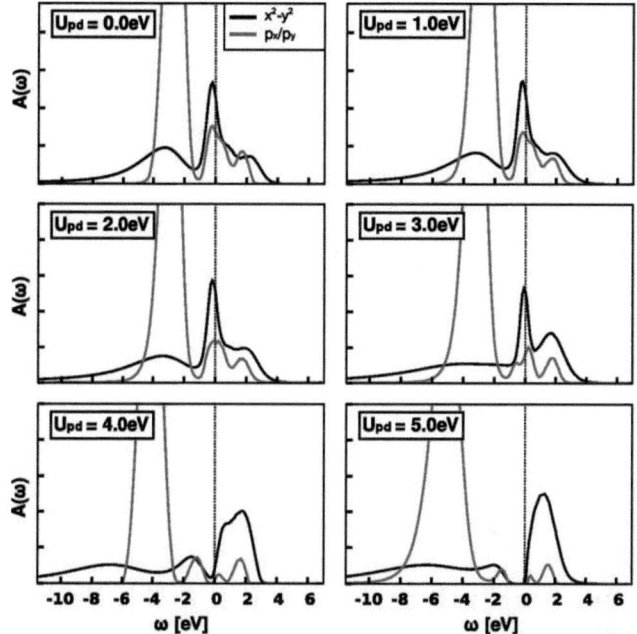

Fig. 5.8: LDA+HartreeDMFT spectral functions for the La$_2$CuO$_4$ (N=0)MTO model for $U_{dd} = 10$ eV, $U_{pp} = 5$ eV and six different values of $U_{pd} \leqq U_{pp}$. For $U_{pd} \gtrsim 4$ eV HartreeDMFT yields an insulating state for the original (N=0)MTO parameters *without* the need to artificially enhance the d–p splitting [103].

5 Expanding the basis set

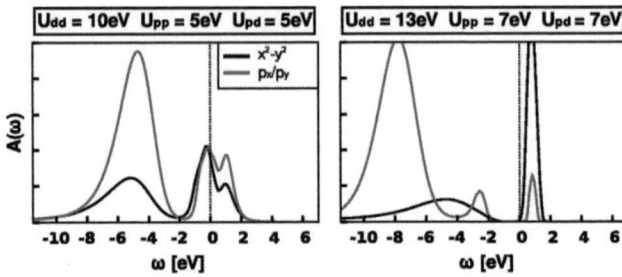

Fig. 5.9: LDA+HartreeDMFT spectral functions for the La_2CuO_4 (N=1)MTO model for $U_{dd} = 10$ eV, $U_{pp} = 5$ eV, and $U_{pd} = 5$ eV (left hand side) and $U_{dd} = 13$ eV, $U_{pp} = 7$ eV, and $U_{pd} = 7$ eV. For the La_2CuO_4 (N=1)MTO model only unrealistically large values of the interaction parameters yield an insulating solution.

ment with experimental evidence of a spectral gap in undoped cuprates in a physically reasonable parameter range.

Conclusion Let us stress again that our main goal was to perform calculations based on the original NMTO parameters *without artificially changing them*. We have seen that, if orbital overlaps are strong, the assumption of an interorbital interaction U_{pd} can be extremely important, since it yields a *self consistently determined level splitting*.

Nonetheless, the results for the N=1 model showed that for realistic parameters the half filled antiferromagnetic insulating phase cannot be comprehended by a single site DMFT or even HartreeDMFT approach. Based on this observation we speculate that it might well be, that non–local correlation effects beyond the reach of DMFT or HartreeDMFT play an important role in the cuprates due to their strong anisotropy. The idea that the insulating phase of the undoped cuprates *cannot be described completely in terms of a strong coupling* system was recently also put forward by Comanac et al. [40] who analyzed experimental data for the optical conductivity of La_2CuO_4.

5.2 Multilayer DMFT

The final project we will address in this work leads us to an extension of the single site DMFT intended for calculations of strongly correlated multi-layered heterostructures.

heterostructures in experiment and theory   As we already mentioned in chapter 4, in recent years there has been a growing interest in the electronic properties of surfaces and interfaces of strongly correlated materials[45]. For many compounds experimentally obtained photoemission spectra no longer agree with theoretically derived bulk spectra of a number of transition-metal oxides. This discrepancy has been attributed to changes in the electronic structure at the surface of these materials [123, 119, 185, 186]. It was suggested to exploit such effects and to use the interfaces of heterostructures as novel correlated devices [146, 194, 61, 117, 212, 111]. Also within this work (chapter 4) we have seen that material engineering of transition metal oxides bears promising opportunities. For instance, the probably best-known example for remarkable heterostructure physics is the interface between $LaTiO_3$ and $SrTiO_3$, which exhibits metallic behavior in spite of the fact that the two constituent bulk materials are insulators [146].

Also on the theoretical side, inhomogeneous strongly correlated layered systems have been subject to studies within the DMFT approach. Potthoff and Nolting [158, 160, 159] investigated a surface metal–insulator transition of the single-band Hubbard model, i.e. an interface to vacuum. Within a three-band model for t_{2g} valence bands of cubic transition metal oxides Liebsch [113] and Ishida [91] showed that surface electrons experience stronger correlation effects due to a delicate interplay of level-splittings, hopping amplitudes, and coulombic interactions. Further single-band Hubbard model studies by Helmes et al. [84] and Okamoto and Millis [147, 148, 149] considered strongly correlated Mott insulators "sandwiched" by either metallic or *band*-insulating layers. Analogous calculations for heterostructures con-

5 Expanding the basis set

sisting of correlated model systems were also carried out by Kancharla and Dagotto [99] and Rüegg et al. [176].
Assuming inhomogeneities or semi–infinite (e.g. surfaces or interfaces) systems breaks the symmetry in the direction normal to the layer planes and presents an additional challenge solving such systems, which is frequently done by employing finite size[5] models: Freericks [63, 38, 64] solved the DMFT equation using the Falikov–Kimball Hamiltonian for a doped Mott insulator sandwiched between two semi–infinite metals, considering 30 layers in each of the metallic leads and up to 80 layers in the barrier region explicitly. Ishida and Liebsch [90] recently suggested a so called tight binding embedding approach to handle a finite number of correlated layers between substrates which, themselves can be correlated.

the layer DMFT scheme In our work we will, however, as a first step assume periodic stacks of layers so that the symmetry of the system is *not broken* if we simply enlarge our unit cell. In this case we can still define $k_\perp$ perpendicular to the layer planes as a good quantum number. Such a situation is for example realized in the case of the nickelate–based 1/1 stacks $LaAlO_3/LaNiO_3$ we discussed in chapter 4.
In the case of these nickelates though, the insulating $LaAlO_3$ layers were designed in order to *confine* the $LaNiO_3$ e_g bands and yield a quasi two dimensional system. Hence, we were able to perform LDA+DMFT calculations without taking into account the $LaAlO_3$ layers explicitly, since excitations connected to these degrees of freedom were far above the energy scale that we are interested in.
This, however, does not always have to be the case. It depends on the composition of the "sandwiched" materials. In order to also account for a more general situation of interplay between different layers, the following self consistent solution for periodic stacks could be performed: The onsite interaction of each layer should be treated with the DMFT style mapping on an impurity model, but *all layers* should be solved simultaneously and, most

---
[5] i.e. finite number of layers along the z–direction

5.2 Multilayer DMFT

importantly, they should be allowed to "communicate" via hopping or interactions within the self consistent iterations.

Like in the previous section, we follow again the philosophy of embedding a single site DMFT iteration in a self consistent loop to solve a problem on a large basis set. Hence, also for the layer DMFT we employ projections on local subspaces (first the layers and then the impurity sites within the layers) of the Green function and perform DMFT calculations within these subspaces. Let us line out the procedure in the same schematic way used before:

1. As initialization we have to take care of the double counting corrections and choose a starting self energy. Since, in general, we would like to perform a full HartreeDMFT for each layer, we again separate the self energy into its components: The full self energy is the sum of a *static* Σ^H and a *dynamic* part Σ^DMFT. Additionally the self energy and also the Green function get a layer index since we want to define a different kinetic energy and interaction strength on each layer (roman numerals in the following discussion). In order to calculate the *full* Green function we integrate over k:

$$\underline{\underline{G}}^\mathrm{loc.}_\mathrm{full}(\omega) = \frac{1}{V_\mathrm{BZ}} \int_\mathrm{BZ} d^3k \left[(\omega + \mu)\underline{\underline{\mathbb{1}}} - \underline{\underline{\varepsilon}}(\mathbf{k}) - \underline{\underline{\Sigma}}^\mathrm{full}(\omega) \right]^{-1} \quad (5.13)$$

where the double lines now denote not only spin and orbital degrees of freedom, but also incorporate the layer index ℓ. Hence, the dispersion has the form

$$\begin{pmatrix} \underline{\varepsilon}_1(\mathbf{k}) & \cdots & & \underline{\varepsilon}_{N,1}(\mathbf{k}) \\ \vdots & \ddots & & \vdots \\ \underline{\varepsilon}_{1,\ell}(\mathbf{k}) & & \underline{\varepsilon}_\ell(\mathbf{k}) & & \underline{\varepsilon}_{N,\ell}(\mathbf{k}) \\ \vdots & & & \ddots & \vdots \\ \underline{\varepsilon}_{1,N}(\mathbf{k}) & \cdots & & & \underline{\varepsilon}_N(\mathbf{k}) \end{pmatrix} \quad (5.14)$$

and we assume the chemical potential μ to be the same for all layers.

5 Expanding the basis set

2. Next, we separate the local Green function for the layers:

$$\underline{G}_\ell^{\text{loc.}}(\omega) = \mathcal{P}_\ell \, \underline{G}_{\text{full}}^{\text{loc.}}(\omega) \qquad (5.15)$$

where $\mathcal{P}_\ell$ is the Projector on the subspace of layer "ℓ". As we described already in the HartreeDMFT the information about the other layers (or "Hartree only" orbitals) is encoded, also in this case, in the local Green function due to the inversion in (5.13).

3. Once we have separated the layer Green functions, we construct the Weiss field for each *layer dependent* impurity model whereby the information of the other layers (possibly even interlayer interaction) is included through the Weiss field

$$\left[\underline{\mathcal{G}}_{\ell,\text{d}}^0(i\omega_\nu)\right]^{-1} = \left[\underline{G}_{\ell,\text{d}}^{\text{loc.}}(\omega)\right]^{-1} + \underline{\Sigma}_{\ell,\text{d}}^{\text{DMFT}}(\omega) \qquad (5.16)$$

where $\underline{G}_{\ell,\text{d}}^{\text{loc.}}(\omega) = \mathcal{P}_\text{d} \, \underline{G}_\ell^{\text{loc.}}(\omega)$, i.e., the projection onto the subspace that should be treated with DMFT. We have to subtract only the DMFT part of the self energy on the d–subspace of layer ℓ (not the Hartree part). In other words *we must only subtract what the corresponding solution of the impurity problem gives back to us:* $\underline{\Sigma}_{\ell,\text{d}}^{\text{DMFT}}(\omega)$

4. The rest of the loop is straight forward: after the calculation of a possible Hartree self energy accounting for d–p or even interlayer interaction we assemble the total self energy and start over until we are converged.

In Fig. 5.10 we show a visualization of the layer DMFT scheme flow in a diagrammatic way.

5.2 Multilayer DMFT

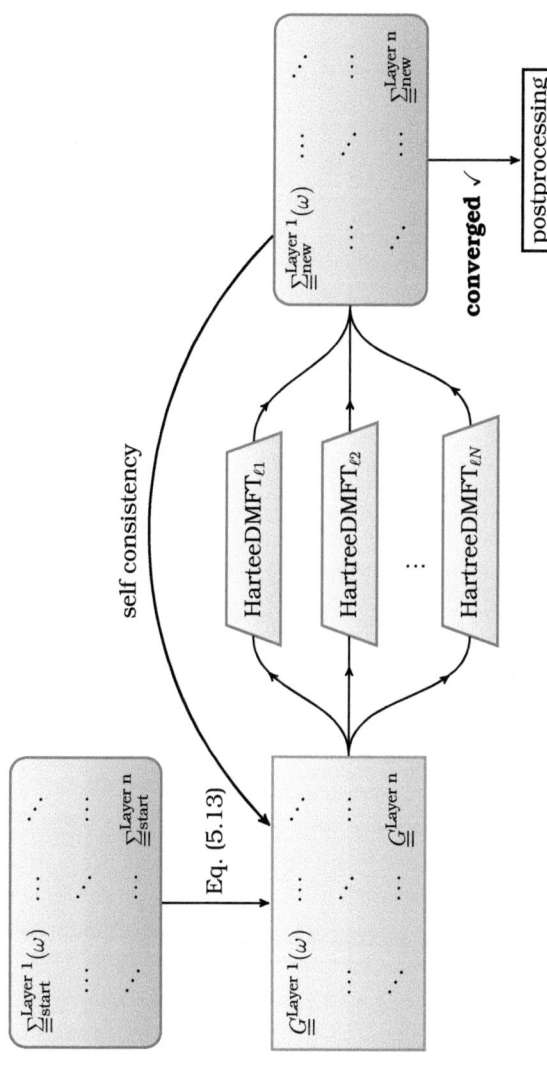

Fig. 5.10: Scheme of the multiple layer DMFT. Single site DMFT iterations for N layers are embedded in a self consistent loop to solve the full layered structure problem on a large basis set. We employ projections first on the layer subspaces and then, within each layer on the impurity sites.

5 Expanding the basis set

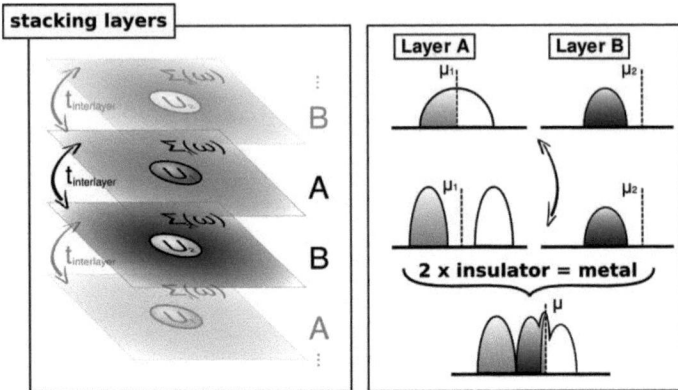

Fig. 5.11: Left hand side: Cartoon of a periodic ...ABABA... stack of layers. Each layer may have different intralayer hoppings (light and dark gray) and local interaction parameters $U_A \neq U_B$. Moreover, we consider the interlayer hopping t_{inter}. Right hand side: schematic sketch of the layer spectral functions. As bulk materials the compound A and B may have different chemical potentials, i.e., different filling. We consider compound A to be half–filled while compound B should have a full band below its Fermi energy, i.e., it is a band insulator (first row). Furthermore, we consider compound A to be a strongly correlated Mott insulator (second row). Stacking those two insulators in a layered structure turns out to yield a metallic solution due to interlayer doping as we sketch it in the bottom row

5.2.1 periodic stacks

Let us now take an example in order to illustrate how the layer DMFT scheme works in our implementation. A good and simple toy model is a periodic stack of layers in a "...ABABAB..." fashion. In Fig. 5.11 on the left hand side we show a sketch of this situation. Moreover let us assume that the compound A (light gray) would be a strongly correlated Mott insulator ($U_A > U_{\text{crit.}}$) in the bulk as it is sketched on the right hand side of the figure. Compound B, on the other hand, we assume to be a band insulator as a bulk compound. Further, for the layers we assume a two–dimensional cubic dispersion relations with intralayer hoppings of $t^A_{\text{intra}} = t^B_{\text{intra}} = 0.25$ eV which leads to a bandwidth

5.2 Multilayer DMFT

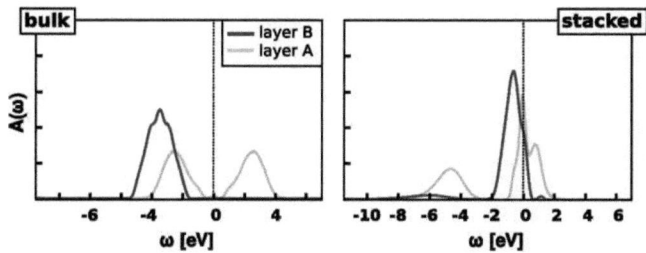

Fig. 5.12: Layer DMFT spectral functions. Left hand side: DMFT results for the bulk systems of the Mott insulating compound A and the band insulating compound B (compare sketch in Fig. 5.11)

of 2 eV for each (isolated) layer. As interaction parameters we assume for both layers $U^{A/B} = 5$ eV, which is chosen to be larger than the critical U value for the Mott transition in the bulk compound A.

Now we bring the two compounds together in a 1/1 layer stack (see Fig. 5.11) and allow for an inter–layer hybridization $\varepsilon_{\text{interl.}} = 2t_{\text{inter}}\cos(k_z)$ (as in Eq. (5.14)) with $t_{\text{inter}} = 0.1$ eV. The chemical potential of such heterostructure is fixed by the filling of the total system which yields three electrons per unit cell. Moreover, we include a difference in the onsite energy of layer B with respect to layer A of $\varepsilon_A - \varepsilon_B = \Delta E = 1.5$ eV. Let us remark that in a realistic case the value of ΔE which will depend on the details of the structure, as well as the hopping parameters should be extracted, from an *ab initio* LDA calculation. In Fig. 5.12 we show on the left hand side the DMFT spectral functions of compound A and compound B in the bulk which we find to be Mott– and band– insulating respectively. On the right hand side we show the result of our layer DMFT calculations for the spectrum of the periodic stack of layers ... ABABA ... which shows instead clear metallic behavior. Note that the coherent quasiparticle excitations around the Fermi energy ε_F have a mixed character, i.e., *both* layers become metallic in the stacked configuration. This insulator to metal transition from bulk to heterostructure can be understood as a doping due to "inter–layer charge exchange": In the stacked structure the charge is arranged differently compared to the bulk compounds mainly due to the differences between the bulk and heterostructure onsite

5 Expanding the basis set

energy. This charge rearrangement can be interpreted as the electron doping of the Mott–insulating A layers or, vice versa, as the hole doping of the band–insulating B layers – in both cases we effectively have a "doping" driven MIT. In spite of the simplicity of this toy model we can comprehend physics which is experimentally observed: The interfaces of $LaTiO_3$ and $SrTiO_3$ [146] as well as $LaAlO_3$ and $SrTiO_3$ [144, 145] show metallic conductivity, although the respective bulk materials are Mott– (LTO) and band–insulating (STO, LAO).

Our new layer DMFT implementation, hence, is capable of treating models which incorporate complex interplay of the constituents in a strongly correlated heterostructure. By means of the projections we capture a large number of important degrees of freedom in the full self consistent solution of the problem similarly to the steps we took in the implementation of the HartreeDMFT. Compared to the HartreeDMFT the layer DMFT implementation is yet "one level higher", since we can now compute a HartreeDMFT for *each* layer which couples to all the other HartreeDMFT layers in the full self consistency of the layer DMFT. Expanding the DMFT basis set even further along these lines would eventually lead us to a self consistency loop as large as the LDA itself, i.e to a full self consistent LDA+DMFT scheme. The implementation of such a scheme is subject to forefront research in the field of *ab initio* calculations for strongly correlated systems allowing finally for a complete treatment of electronic correlations an their feedback effects on the crystal structure. The extensions of the standard DMFT which we presented in this final chapter of the thesis show the path along which we can achive this important goal.

6 Summary & Outlook

The main topic of this work was i) the improvement of understanding correlated electron systems, ii) the study of how new knowledge can be exploited in order to *design* solids to suit our needs, and iii) the development of new methods in order to provide a sound theoretical fundament. These aspects are reflected in the three chapters following the pedagogical introduction.

In chapter 3 we encountered two well known correlated materials, vanadium sesquioxide V_2O_3 and nickel disulfide NiS_2. In order to better understand the non–trivial ground states of these systems, we analyzed carefully the metal to insulator transitions driven by external parameters like pressure and doping. For *both* systems we were able to theoretically interpret experimental data, proving previous simplifying assumptions concerning the microscopic mechanisms (pressure ↔ doping equivalence) to be incorrect.
For pure and Cr doped V_2O_3 we performed a novel combined theoretical analysis of state–of–the–art LDA+DMFT together with full–multiplet configuration–interaction cluster calculations for the interpretation of x–ray absorption spectra on the vanadium K–edge. In this way, we were able to find a robust probe for the ground state which could be employed also for high pressure experiments. Previous probe techniques (on the V–L edge) were not applicable in the high pressure regime, so that our results filled an important piece in the ground state puzzle of V_2O_3. Furthermore, we analyzed the region in the phase diagram close to the Mott transition and confirmed a phase mixture scenario which was previously only speculative. In future projects, we plan to study the modeling of mixed state phases by means of DMFT schemes which are capable of including disorder potentials.

6 Summary & Outlook

Chapter 4 was devoted to exploiting the knowledge the cuprate superconductors in order to find candidates for non–cuprate high temperature superconductivity. In that respect we investigated nickel–based "bulk" and artificial "heterostructures". For the bulk systems we have shown that previously discarded compounds, namely the $LnSrNiO_4$ series, have a potential for cuprate–like superconductivity which may have been underestimated. In the heterostructure compound $LaAlO_3/LaNiO_3$, we find beside antiferromagnetic fluctuations, a Fermi surface structure very similar to the one of the cuprates. Hence, the basic ingredients of high T_C superconductivity according to many suggested scenarios. This work triggered several experimental efforts to realize nickel–based high T_C superconductors. Collaboration with the experimental groups yields a large spectrum of possible studies.

The last chapter focused on the methodological advances of this work. We implemented the HartreeDMFT scheme in order to efficiently treat systems where the minimal low energy models allow for a separation into a correlated subspace, treated with DMFT, and the remaining degrees of freedom which can be treated more approximative by static Hartree mean field theory. With this novel scheme we revisited the Emery–like three band models for La_2CuO_4, and carefully analyzed two different NMTO *ab initio* models. From our study we concluded that artificially introduced parameter changes relative to the *ab initio* derived NMTO models of previous studies could be avoided by taking into account an additional interaction between oxygen p– and copper d–states. Moreover, our study yields indication, that the antiferromagnetic insulating ground state of La_2CuO_4 is not a simple local U Mott–insulating state but possibly connected to longer range correlations beyond the single site DMFT approach.

In the second part of the last chapter we introduced a DMFT scheme for the calculation of layered heterostructures. The implementation was motivated by fast developing synthesis technology. With the calculation of a toy model for a periodically stacked compound we showed, how dramatic the interplay of strongly correlated materials in a heterostructure changes the physics with respect to the bulk compounds of its constituents.

Bibliography

[1] A. Abrikosov, L. Gorkov, and Dzyaloshinski, *Methods of Quantum Field Theory in Statistical Physics*, Dover (1975)

[2] O. K. Andersen, *Linear methods in band theory*, Phys. Rev. B, **12**(8), 3060 (1975)

[3] O. K. Andersen, *Electronic Structure and Physical Properties of Solids: The Uses of the LMTO method*, Lecture Notes in Physics Springer, New York, 2000 (2000)

[4] O. K. Andersen and O. Jepsen, *Explicit, First–Principles Tight–Binding Theory*, Phys. Rev. Lett., **53**, 2571 (1984)

[5] O. K. Andersen, A. I. Liechtenstein, O. Jepsen, and F. Paulsen, *LDA energy bands, low–energy hamiltonians, t', t'', $t_{perp.}(k)$, and $J_{perp.}$*, Journal of Physics and Chemistry of Solids, **56**, 1573 (1995)

[6] O. K. Andersen and T. Saha-Dasgupta, *Muffin–tin orbitals of arbitrary order*, Phys. Rev. B, **62**, R16219 (2000)

[7] V. Anisimov, I. Nekrasov, D. Kondakov, T. Rice, and M. Sigrist, *Orbital-selective Mott–insulator transition in $Ca_{2-x}Sr_xRuO_4$*, The European Physical Journal B - Condensed Matter and Complex Systems, **25**, 191 (2002)

[8] V. I. Anisimov, D. Bukhvalov, and T. M. Rice, *Electronic structure of possible nickelate analogs to the cuprates*, Phys. Rev. B, **59**, 7901 (1999)

[9] V. I. Anisimov, J. Zaanen, and O. K. Andersen, *Band theory and Mott insulators: Hubbard U instead of Stoner I*, Phys. Rev. B, **44**, 943 (1991)

[10] E. Arcangeletti, L. Baldassarre, D. D. Castro, S. Lupi, L. Malavasi, C. Marini, A. Perucchi, and P. Postorino, *Evidence of a Pressure-Induced Metallization Process in Monoclinic VO_2*, Physical Review Letters, **98**, 196406 (2007)

[11] R. Arita and K. Held, *Orbital-selective Mott-Hubbard transition in the two-band Hubbard model*, Phys. Rev. B, **72**, 201102 (2005)

[12] Ashcroft and Mermin, *Solid State Physics*, Hartcourt College Publishers, New York (1976)

[13] L. Baldassarre, A. Perucchi, E. Arcangeletti, D. Nicoletti, D. D. Castro, P. Postorino, V. A. Sidorov, and S. Lupi, *Electrodynamics near the metal-to-insulator transition in V_3O_5*, Physical Review B (Condensed Matter and Materials Physics), **75**, 245108 (2007)

[14] L. Baldassarre, A. Perucchi, D. Nicoletti, A. Toschi, G. Sangiovanni, K. Held, M. Capone, M. Ortolani, L. Malavasi, M. Marsi, P. Metcalf, P. Postorino, and S. Lupi, *Quasiparticle evolution and pseudogap formation in V_2O_3: An infrared spectroscopy study*, Physical Review B (Condensed Matter and Materials Physics), **77**, 113107 (2008)

[15] J. G. Bednorz and K. A. Müller, *Possible high T_C superconductivity in the Ba–La–Cu–O system*, Zeitschrift für Physik B Condensed Matter, **64**, 189 (1986)

[16] M. A. Beno, L. Soderholm, D. W. C. II, D. G. Hinks, J. D. Jorgensen,

J. D. Grace, I. K. Schuller, C. U. Segre, and K. Zhang, *Structure of the single-phase high-temperature superconductor $YBa_2Cu_3O_{7-\delta}$*, Applied Physics Letters, **51**(1), 57 (1987)

[17] A. Bianconi and C. R. Natoli, *Effect of the metal-insulator transition on vanadium K-photoabsortion spectrum in V_2O_3*, Solid State Communications, **27**(11), 1177 (1978)

[18] N. Blümer, *Efficiency of quantum Monte Carlo impurity solvers for the dynamical mean-field theory*, Phys. Rev. B, **76**, 205120 (2007)

[19] W. F. Brinkman and T. M. Rice, *Application of Gutzwiller's Variational Method to the Metal-Insulator Transition*, Phys. Rev. B, **2**(10), 4302 (1970)

[20] L. N. Bulaevskii, E. L. Nagaev, and D. L. Khomskii, Sov. Phys. JETP, **27**, 838 (1968)

[21] R. Bulla, *Zero Temperature Metal-Insulator Transition in the Infinite-Dimensional Hubbard Model*, Phys. Rev. Lett., **83**(1), 136 (1999)

[22] R. Bulla, A. C. Hewson, and T. Pruschke, *Numerical renormalization group calculations for the self energy of the impurity Anderson model*, J. Phys.: Condens. Matter, **10**, 8365 (1998)

[23] M. Caffarel and W. Krauth, *Exact diagonalization approach to correlated fermions in infinite dimensions: Mott transition and superconductivity*, Phys. Rev. Lett., **72**(10), 1545 (1994)

[24] E. Cappelluti and S. Ciuchi, *Magnetic and lattice polaron in the Holstein $t-J$ model*, Phys. Rev. B, **66**(16), 165102 (2002)

[25] G. L. Carr, S. Perkowitz, and D. B. Tanner, *Infrared and milimeter*

waves, Academic Press, Orlando (1985)

[26] P. Carra, B. T. Thole, M. Altarelli, and X. Wang, *X-ray circular dichroism and local magnetic fields*, Phys. Rev. Lett., **70**(5), 694 (1993)

[27] C. Castellani, C. R. Natoli, and J. Ranninger, *Insulating phase of V_2O_3: An attempt at a realistic calculation*, Phys. Rev. B, **18**, 4967 (1978)

[28] C. Castellani, C. R. Natoli, and J. Ranninger, *Magnetic structure of V_2O_3 in the insulating phase*, Phys. Rev. B, **18**, 4945 (1978)

[29] C. Castellani, C. R. Natoli, and J. Ranninger, *Metal-insulator transition in pure and Cr-doped V_2O_3*, Phys. Rev. B, **18**, 5001 (1978)

[30] R. J. Cava, A. Santoro, D. W. Johnson, and W. W. Rhodes, *Crystal structure of the high-temperature superconductor $La_{1.85}Sr_{0.15}CuO_4$ above and below T_C*, Phys. Rev. B, **35**(13), 6716 (1987)

[31] R. J. Cava *et al.*, Physica (Amsterdam), **172 C**, 138 (1990)

[32] D. M. Ceperley and B. J. Alder, *Ground State of the Electron Gas by a Stochastic Method*, Phys. Rev. Lett., **45**(7), 566 (1980)

[33] J. Chakhalian, J. W. Freeland, H.-U. Habermeier, G. Cristiani, G. Khaliullin, M. van Veenendaal, and B. Keimer, *Orbital Reconstruction and Covalent Bonding at an Oxide Interface*, Science, **318**, 1114 (2007)

[34] J. Chaloupka and G. Khaliullin, *Orbital Order and Possible Superconductivity in $LaNiO_3/LaMO_3$ Superlattices*, Physical Review Letters, **100**, 016404 (2008)

[35] C. T. Chen and F. Sette, *High Resolution Soft X-Ray Spectroscopies with the Dragon Beamline*, Physica Scripta, **T31**, 119 (1990)

[36] C. T. Chen, F. Sette, Y. Ma, and S. Modesti, *Soft-x-ray magnetic circular dichroism at the L2,3 edges of nickel*, Phys. Rev. B, **42**(11), 7262 (1990)

[37] C. T. Chen, L. H. Tjeng, J. Kwo, H. L. Kao, P. Rudolf, F. Sette, and R. M. Fleming, *Out-of-plane orbital characters of intrinsic and doped holes in $La_{2-x}Sr_xCuO_4$*, Phys. Rev. Lett., **68**(16), 2543 (1992)

[38] L. Chen and J. K. Freericks, *Electronic charge reconstruction of doped Mott insulators in multilayered nanostructures*, Phys. Rev. B, **75**, 125114 (2007)

[39] O. Chmaissem, J. D. Jorgensen, S. Short, A. Knizhnik, Y. Eckstein, and H. Shaked, *Scaling of transition temperature and CuO_2 plane buckling in a high-temperature superconductor*, Nature, **397**(6714), 45 (1999)

[40] A. Comanac, L. de/' Medici, M. Capone, and A. J. Millis, *Optical conductivity and the correlation strength of high-temperature copper-oxide superconductors*, Nat Phys, **4**(4), 287 (2008)

[41] A. Congeduti, P. Postorino, P. Dore, A. Nucara, S. Lupi, S. Mercone, P. Calvani, A. Kumar, and D. D. Sarma, *Infrared study of charge delocalization induced by pressure in the $La_{0.75}Ca_{0.25}MnO_3$ manganite*, Phys. Rev. B, **63**, 184410 (2001)

[42] K. D. Cummings, J. C. Garland, and D. B. Tanner, *Optical properties of a small-particle composite*, Phys. Rev. B, **30**, 4170 (1984)

[43] G. Czycholl, *Theoretische Festkörperphysik*, Springer-Verlag GmbH (2007)

[44] E. Dagotto, *Complexity in Strongly Correlated Electronic Systems*, Science, **309**(5732), 257 (2005)

[45] E. Dagotto, *When Oxides Meet Face to Face*, Science, **318**(5853), 1076 (2007)

[46] H. Das and T. Saha-Dasgupta, *Electronic structure of La_2CuO_4 in the T and T' crystal structures using dynamical mean field theory*, Phys. Rev. B, **79**, 134522 (2009)

[47] F. de Groot, *High-Resolution X-ray Emission and X-ray Absorption Spectroscopy*, Chemical Reviews, **101**(6), 1779 (2001)

[48] F. M. F. de Groot, *X–ray absorption and dichroism of transition metals and their compounds*, Journal of Electron Spectroscopy and Related Phenomena, **67**(4), 529 (1994)

[49] L. de'Medici, A. Georges, and S. Biermann, *Orbital-selective Mott transition in multiband systems: Slave-spin representation and dynamical mean-field theory*, Phys. Rev. B, **72**(20), 205124 (2005)

[50] S. Di Matteo, N. B. Perkins, and C. R. Natoli, *Spin-1 effective Hamiltonian with three degenerate orbitals: An application to the case of V_2O_3*, Phys. Rev. B, **65**, 054413 (2002)

[51] M. Dressel and G. Grüner, *Electrodynamics of Solids*, Cambridge University Press, Cambridge, England (2002)

[52] I. S. Elfimov, T. Saha-Dasgupta, and M. A. Korotin, *Role of c-axis pairs in V_2O_3 from the band-structure point of view*, Phys. Rev. B, **68**, 113105 (2003)

[53] I. S. Elfimov, N. A. Skorikov, V. I. Anisimov, and G. A. Sawatzky, *Band Structure Approach to Resonant X-Ray Scattering*, Phys. Rev. Lett., **88**, 015504 (2001)

[54] V. J. Emery, *Theory of high–T_C superconductivity in oxides*, Phys. Rev. Lett., **58**, 2794 (1987)

[55] V. J. Emery and S. A. Kivelson, *Importance of phase fluctuations in superconductors with small superfluid density*, Nature, **374**(6521), 434 (1995)

[56] S. Y. Ezhov, V. I. Anisimov, D. I. Khomskii, and G. A. Sawatzky, *Orbital Occupation, Local Spin, and Exchange Interactions in V_2O_3*, Phys. Rev. Lett., **83**, 4136 (1999)

[57] L. M. Falicov and J. C. Kimball, *Simple Model for Semiconductor-Metal Transitions: SmB_6 and Transition-Metal Oxides*, Phys. Rev. Lett., **22**(19), 997 (1969)

[58] M. Feldbacher, K. Held, and F. F. Assaad, *Projective Quantum Monte Carlo Method for the Anderson Impurity Model and its Application to Dynamical Mean Field Theory*, Phys. Rev. Lett., **93**, 136405 (2004)

[59] M. Ferrero, F. Becca, M. Fabrizio, and M. Capone, *Dynamical behavior across the Mott transition of two bands with different bandwidths*, Phys. Rev. B, **72**(20), 205126 (2005)

[60] J. Fink, T. Müller-Heinzerling, B. Scheerer, W. Speier, F. U. Hillebrecht, J. C. Fuggle, J. Zaanen, and G. A. Sawatzky, *2p absorption spectra of the 3d elements*, Phys. Rev. B, **32**(8), 4899 (1985)

[61] L. Fitting Kourkoutis, Y. Hotta, T. Susaki, H. Y. Hwang, and D. A. Muller, *Nanometer Scale Electronic Reconstruction at the Interface between $LaVO_3$ and $LaVO_4$*, Phys. Rev. Lett., **97**(25), 256803 (2006)

[62] W. Folkerts, G. A. Sawatzky, C. Haas, R. A. de Groot, and F. U. Hille-

brecht, *Electronic structure of some 3D transition-metal pyrites*, Journal of Physics C: Solid State Physics, **20**(26), 4135 (1987)

[63] J. K. Freericks, *Dynamical mean-field theory for strongly correlated inhomogeneous multilayered nanostructures*, Phys. Rev. B, **70**(19), 195342 (2004)

[64] J. K. Freericks, V. Zlatić, and A. M. Shvaika, *Electronic thermal transport in strongly correlated multilayered nanostructures*, Phys. Rev. B, **75**, 035133 (2007)

[65] A. I. Frenkel, D. M. Pease, J. I. Budnick, P. Metcalf, E. A. Stern, P. Shanthakumar, and T. Huang, *Strain-Induced Bond Buckling and Its Role in Insulating Properties of Cr-Doped V_2O_3*, Phys. Rev. Lett., **97**, 195502 (2006)

[66] A. Fujimori, *Pressure Effects on Transition-Metal Compounds near Insulator-Metal Phase Boundaries*, Phys. Status Solidi B, **233**(1), 47 (2001)

[67] A. Georges and G. Kotliar, *Hubbard model in infinite dimensions*, Phys. Rev. B, **45**, 6479 (1992)

[68] A. Georges, G. Kotliar, W. Krauth, and M. J. Rozenberg, *Dynamical mean-field theory of strongly correlated fermion systems and the limit of infinite dimensions*, Rev. Mod. Phys., **68**, 13 (1996)

[69] A. Georges and W. Krauth, *Numerical solution of the $d = \infty$ Hubbard model: Evidence for a Mott transition*, Phys. Rev. Lett., **69**(8), 1240 (1992)

[70] D. Goldschmidt, G. M. Reisner, Y. Direktovitch, A. Knizhnik, E. Gartstein, G. Kimmel, and Y. Eckstein, *Tetragonal superconducting fam-*

ily (Ca_xLa_{1-x}) $(Ba_{1.75-x}La_{0.25+x})Cu_3O_y$: The effect of cosubstitution on the transition temperature, Phys. Rev. B, **48**(1), 532 (1993)

[71] C. Gougoussis, M. Calandra, A. Seitsonen, C. Brouder, A. Shukla, and F. Mauri, Intrinsic charge transfer gap in NiO from Ni K–edge x-ray absorption spectroscopy, Phys. Rev. B, **79**(4), 045118 (2009)

[72] J. Goulon, A. Rogalev, C. Goulon-Ginet, G. Benayoun, L. Paolasini, C. Brouder, C. Malgrange, and P. A. Metcalf, First Observation of Non-reciprocal x–ray Gyrotropy, Phys. Rev. Lett., **85**, 4385 (2000)

[73] P. Hansmann, Crystal-Field ground states of rare-earth materials determined by linear dichroism: A feasibility study and its experimental proof, Master's thesis, University of Cologne (2007)

[74] P. Hansmann, A. Severing, Z. Hu, M. W. Haverkort, C. F. Chang, S. Klein, A. Tanaka, H. H. Hsieh, H.-J. Lin, C. T. Chen, B. Fåk, P. Lejay, and L. H. Tjeng, Determining the Crystal-Field Ground State in Rare Earth Heavy Fermion Materials Using Soft-X-Ray Absorption Spectroscopy, Phys. Rev. Lett., **100**(6), 066405 (2008)

[75] P. Hansmann, X. Yang, A. Toschi, G. Khaliullin, O. K. Andersen, and K. Held, Turning a Nickelate Fermi Surface into a Cupratelike One through Heterostructuring, Phys. Rev. Lett., **103**(1), 016401 (2009)

[76] M. W. Haverkort, Spin and orbital degrees of freedom in transition metal oxides and oxide thin films studied by soft x–ray absorbtion spectroscopy, Ph.D. thesis, Universität zu Köln (2005)

[77] M. W. Haverkort, S. I. Csiszar, Z. Hu, S. Altieri, A. Tanaka, H. H. Hsieh, H.-J. Lin, C. T. Chen, T. Hibma, and L. H. Tjeng, Magnetic versus crystal-field linear dichroism in NiO thin films, Phys. Rev. B, **69**(2), 020408 (2004)

Bibliography

[78] M. W. Haverkort, E. Pavarini, L. H. Tjeng, Z. Hu, A. Tanaka, H. H. Hsieh, H.-J. Lin, and C. T. Chen, *Orbital ordering and x-ray dichroism in ReTiO$_3$* (2010), Private communication

[79] L. Hedin and B. Lundqvist, J. Phys. C: Solid State Phys., **4**, 2064 (1971)

[80] K. Held, *Untersuchungen korrelierter Elektronensysteme im Rahmen der Dynamischen Molekularfeldtheorie*, Ph.D. thesis, Universitat Augsburg (1999)

[81] K. Held, *Electronic structure calculations using dynamical mean field theory*, Advances in Physics, **56**, 829 (2007)

[82] K. Held, G. Keller, V. Eyert, D. Vollhardt, and V. I. Anisimov, *Mott-Hubbard Metal-Insulator Transition in Paramagnetic V$_2$O$_3$: A LDA+DMFT(QMC) Study*, Phys. Rev. Lett., **86**, 5345 (2001)

[83] K. Held, A. K. McMahan, and R. T. Scalettar, *Cerium Volume Collapse: Results from the Merger of Dynamical Mean-Field Theory and Local Density Approximation*, Phys. Rev. Lett., **87**, 276404 (2001)

[84] R. W. Helmes, T. A. Costi, and A. Rosch, *Kondo Proximity Effect: How Does a Metal Penetrate into a Mott Insulator?*, Phys. Rev. Lett., **101**(6), 066802 (2008)

[85] J. E. Hirsch and R. M. Fye, *Monte Carlo Method for Magnetic Impurities in Metals*, Phys. Rev. Lett., **56**, 2521 (1986)

[86] P. Hohenberg and W. Kohn, *Inhomogeneous Electron Gas*, Phys. Rev., **136**(3B), B864 (1964)

[87] J. Hubbard, *Electron Correlations in Narrow Energy Bands*, Proceed-

ings of the Royal Society of London. Series A, Mathematical and Physical Sciences, **276**, 238 (1963)

[88] B. A. Hunter *et al.*, Physica (Amsterdam), **221 C**, 1 (1994)

[89] M. Imada, A. Fujimori, and Y. Tokura, *Metal-insulator transitions*, Rev. Mod. Phys., **70**, 1039 (1998)

[90] H. Ishida and A. Liebsch, *Embedding approach for dynamical mean-field theory of strongly correlated heterostructures*, Phys. Rev. B, **79**, 045130 (2009)

[91] H. Ishida, D. Wortmann, and A. Liebsch, *Electronic structure of SrVO$_3$ (001) surfaces: A local-density approximation plus dynamical mean-field theory calculation*, Phys. Rev. B, **73**(24), 245421 (2006)

[92] M. Jarrell, *Hubbard model in infinite dimensions: A quantum Monte Carlo study*, Phys. Rev. Lett., **69**, 168 (1992)

[93] M. Jarrell, *Hubbard model in infinite dimensions: A quantum Monte Carlo study*, Phys. Rev. Lett., **69**(1), 168 (1992)

[94] M. Jarrell and J. E. Gubernatis, *Bayesian inference and the analytic continuation of imaginary-time quantum Monte Carlo data*, Physics Reports, **269**(3), 133 (1996)

[95] M. Jarrell and T. Pruschke, *Magnetic and dynamic properties of the Hubbard model in infinite diomensions*, Zeitschrift für Physik B Condensed Matter, **90**, 187 (1993)

[96] M. Jarrell and T. Pruschke, *Anomalous properties of the Hubbard model in infinite dimensions*, Phys. Rev. B, **49**(2), 1458 (1994)

[97] S. Jiuxun, W. Qiang, C. Lingcang, and J. Fuqian, *Two universal equations of state for solids satisfying the limiting condition at high pressure*, Journal of Physics and Chemistry of Solids, **66**, 773 (2005)

[98] R. O. Jones and O. Gunnarsson, *The density functional formalism, its applications and prospects*, Rev. Mod. Phys., **61**(3), 689 (1989)

[99] S. S. Kancharla and E. Dagotto, *Metallic interface at the boundary between band and Mott insulators*, Phys. Rev. B, **74**, 195427 (2006)

[100] A. A. Katanin, A. Toschi, and K. Held, *Comparing pertinent effects of antiferromagnetic fluctuations in the two- and three-dimensional Hubbard model*, Phys. Rev. B, **80**, 075104 (2009)

[101] R. L. Kautz, M. S. Dresselhaus, D. Adler, and A. Linz, *Electrical and Optical Properties of NiS_2*, Phys. Rev. B, **6**, 2078 (1972)

[102] G. Keller, K. Held, V. Eyert, D. Vollhardt, and V. I. Anisimov, *Electronic structure of paramagnetic V_2O_3 : Strongly correlated metallic and Mott insulating phase*, Phys. Rev. B, **70**(20), 205116 (2004)

[103] P. R. C. Kent, T. Saha-Dasgupta, O. Jepsen, O. K. Andersen, A. Macridin, T. A. Maier, M. Jarrell, and T. C. Schulthess, *Combined density functional and dynamical cluster quantum Monte Carlo calculations of the three-band Hubbard model for hole-doped cuprate superconductors*, Phys. Rev. B, **78**, 035132 (2008)

[104] A. Koga, N. Kawakami, T. M. Rice, and M. Sigrist, *Spin, charge, and orbital fluctuations in a multiorbital Mott insulator*, Phys. Rev. B, **72**(4), 045128 (2005)

[105] G. Kotliar, S. Y. Savrasov, K. Haule, V. S. Oudovenko, O. Parcollet,

and C. A. Marianetti, *Electronic structure calculations with dynamical mean-field theory*, Reviews of Modern Physics, **78**, 865 (2006)

[106] G. Kresse and J. Furthmüller, *Efficient iterative schemes for ab initio total-energy calculations using a plane-wave basis set*, Phys. Rev. B, **54**, 11169 (1996)

[107] J. Kuneš, L. Baldassarre, B. Schächner, K. Rabia, C. A. Kuntscher, D. M. Korotin, V. I. Anisimov, J. A. McLeod, E. Z. Kurmaev, and A. Moewes, *Metal-insulator transition in $NiS_{2-x}Se_x$*, Phys. Rev. B, **81**(3), 035122 (2010)

[108] H. Kuwamoto, J. M. Honig, and J. Appel, *Electrical properties of the $(V_{1-x}Cr_x)_2O_3$ system*, Phys. Rev. B, **22**, 2626 (1980)

[109] P. Kwizera, M. S. Dresselhaus, and D. Adler, *Electrical properties of $NiS_{2-x}Se_x$*, Phys. Rev. B, **21**, 2328 (1980)

[110] M. S. Laad, L. Craco, and E. Müller-Hartmann, *Orbital-selective insulator-metal transition in V_2O_3 under external pressure*, Phys. Rev. B, **73**(4), 045109 (2006)

[111] H. N. Lee, H. M. Christen, M. F. Chisholm, C. M. Rouleau, and D. H. Lowndes, *Strong polarization enhancement in asymmetric three-component ferroelectric superlattices*, Nature, **433**(7024), 395 (2005)

[112] T. C. Leung, X. W. Wang, and B. N. Harmon, *Band-theoretical study of magnetism in Sc_2CuO_4*, Phys. Rev. B, **37**(1), 384 (1988)

[113] A. Liebsch, *Surface versus Bulk Coulomb Correlations in Photoemission Spectra of $SrVO_3$ and $CaVO_3$*, Phys. Rev. Lett., **90**(9), 096401 (2003)

[114] P. Limelette, A. Georges, D. Jerome, P. Wzietek, P. Metcalf, and J. M.

Bibliography

Honig, *Universality and Critical Behavior at the Mott Transition*, Science, **302**(5642), 89 (2003)

[115] R. Loudon, *The Quantum Theory of Light*, (Oxford: Clarendon) (1983)

[116] A. R. Mackintosh and O. K. Andersen, *Electrons at the Fermi Surface*, Cambridge University Press, Cambridge (1980)

[117] K. Maekawa, M. Takizawa, H. Wadati, T. Yoshida, A. Fujimori, H. Kumigashira, M. Oshima, Y. Muraoka, Y. Nagao, and Z. Hiroi, *Photoemission study of TiO_2VO_2 interfaces*, Phys. Rev. B, **76**, 115121 (2007)

[118] Y. Maeno, H. Hashimoto, K. Yoshida, S. Nishizaki, T. Fujita, J. G. Bednorz, and F. Lichtenberg, *Superconductivity in a layered perovskite without copper*, Nature, **372**, 532 (1994)

[119] K. Maiti, D. D. Sarma, M. J. Rozenberg, I. H. Inoue, H. Makino, O. Goto, M. Pedio, and R. Cimino, *Electronic structure of $Ca_{1-x}Sr_xVO_3$: A tale of two energy scales*, EPL (Europhysics Letters), **55**(2), 246 (2001)

[120] G. Martinez and P. Horsch, *Spin polarons in the t-J model*, Phys. Rev. B, **44**(1), 317 (1991)

[121] L. F. Mattheiss, *Electronic Structure of the 3d Transition-Metal Monoxides. I. Energy-Band Results*, Phys. Rev. B, **5**(2), 290 (1972)

[122] L. F. Mattheiss, *Band properties of metallic corundum-phase V_2O_3*, Journal of Physics: Condensed Matter, **6**, 6477 (1994)

[123] R. Matzdorf, Z. Fang, Ismail, J. Zhang, T. Kimura, Y. Tokura, K. Terakura, and E. W. Plummer, *Ferromagnetism Stabilized by Lattice Distortion at the Surface of the p-Wave Superconductor Sr_2RuO_4*, Science, **289**(5480), 746 (2000)

[124] D. B. McWhan, A. Menth, J. P. Remeika, W. F. Brinkman, and T. M. Rice, *Metal-Insulator Transitions in Pure and Doped V_2O_3*, Phys. Rev. B, **7**(5), 1920 (1973)

[125] D. B. McWhan and J. P. Remeika, *Metal-Insulator Transition in $(V_{1-x}Cr_x)_2O_3$*, Phys. Rev. B, **2**, 3734 (1970)

[126] D. B. McWhan, T. M. Rice, and J. P. Remeika, *Mott Transition in Cr-Doped V_2O_3*, Phys. Rev. Lett., **23**, 1384 (1969)

[127] T. Mercouris, Y. Komninos, S. Dionissopoulou, and C. A. Nicolaides, *The electric dipole approximation and the calculation of free - free transition matrix elements in multiphoton processes*, Journal of Physics B: Atomic, Molecular and Optical Physics, **30**(9), 2133 (1997)

[128] W. Metzner and D. Vollhardt, *Correlated Lattice Fermions in $d = \infty$ Dimensions*, Phys. Rev. Lett., **62**, 324 (1989)

[129] S. Miyasaka, H. Takagi, Y. Sekine, H. Takahashi, N. Môri, and R. J. Cava, *Metal-Insulator Transition and Itinerant Antiferromagnetism in $NiS_{2-x}Se_x$ Pyrite*, Journal of the Physical Society of Japan, **69**, 3166 (2000)

[130] S.-K. Mo, J. D. Denlinger, H.-D. Kim, J.-H. Park, J. W. Allen, A. Sekiyama, A. Yamasaki, K. Kadono, S. Suga, Y. Saitoh, T. Muro, P. Metcalf, G. Keller, K. Held, V. Eyert, V. I. Anisimov, and D. Vollhardt, *Prominent Quasiparticle Peak in the Photoemission Spectrum of the Metallic Phase of V_2O_3*, Phys. Rev. Lett., **90**, 186403 (2003)

[131] S.-K. Mo, H.-D. Kim, J. D. Denlinger, J. W. Allen, J.-H. Park, A. Sekiyama, A. Yamasaki, S. Suga, Y. Saitoh, T. Muro, and P. Metcalf, *Photoemission study of $(V_{1-x}M_x)_2O_3$ (M=Cr,Ti)*, Phys. Rev. B, **74**, 165101 (2006)

[132] G. Moeller, Q. Si, G. Kotliar, M. Rozenberg, and D. S. Fisher, *Critical Behavior near the Mott Transition in the Hubbard Model*, Phys. Rev. Lett., **74**(11), 2082 (1995)

[133] R. M. Moon, *Antiferromagnetism in V_2O_3*, Phys. Rev. Lett., **25**(8), 527 (1970)

[134] A. A. Mostofi, J. R. Yates, Y.-S. Lee, I. Souza, D. Vanderbilt, and N. Marzari, *Wannier90: A Tool for Obtaining Maximally-Localised Wannier Functions*, Comput. Phys. Commun., **178**, 685 (2008)

[135] N. F. Mott, *The Basis of the Electron Theory of Metals, with Special Reference to the Transition Metals*, Proceedings of the Physical Society. Section A, **62**, 416 (1949)

[136] E. Müller-Hartmann, *Correlated fermions on a lattice in high dimensions*, Zeitschrift für Physik B Condensed Matter, **74**, 507 (1988)

[137] E. Müller-Hartmann, *Fermions on a lattice in high dimensions*, Int. J. Mod. Phys. B, **3**, 2169 (1989)

[138] E. Müller-Hartmann, *The Hubbard model at high dimensions: Some exact results and weak-coupling theory*, Z. Phys. B, **76**, 211 (1989)

[139] F. D. Murnaghan, *The Compressibility of Media under Extreme Pressures*, Proceedings of the National Academy of Sciences of the United States of America, **30**, 244 (1944)

[140] P. G. Niklowitz, M. J. Steiner, G. G. Lonzarich, D. Braithwaite, G. Knebel, J. Flouquet, and J. A. Wilson, *Unconventional resistivity at the border of metallic antiferromagnetism in NiS_2*, arXiv, **06**, 0610166 (2006)

[141] M. D. Núñez Regueiro, M. Altarelli, and C. T. Chen, *X-ray-absorption sum rule for linear dichroism: Application to high-T_C cuprate oxides*, Phys. Rev. B, **51**(1), 629 (1995)

[142] F. J. Ohkawa, J. Phys. Soc. Jap., **61**, 1615 (1991)

[143] F. J. Ohkawa, *Electron correlation in the Hubbard model in $d = \infty$ dimensions*, J. Phys. Soc. Jap., **60**, 3218 (1991)

[144] A. Ohtomo and H. Y. Hwang, *A high-mobility electron gas at the $LaAlO_3/SrTiO_3$ heterointerface*, Nature, **427**(6973), 423 (2004)

[145] A. Ohtomo and H. Y. Hwang, *Corrigendum: A high-mobility electron gas at the $LaAlO_3/SrTiO_3$ heterointerface*, Nature, **441**(7089), 120 (2006)

[146] A. Ohtomo, D. A. Muller, J. L. Grazul, and H. Y. Hwang, *Artificial charge-modulationin atomic-scale perovskite titanate superlattices*, Nature, **419**(6905), 378 (2002)

[147] S. Okamoto and A. J. Millis, *Spatial inhomogeneity and strong correlation physics: A dynamical mean-field study of a model Mott-insulator–band-insulator heterostructure*, Phys. Rev. B, **70**, 241104 (2004)

[148] S. Okamoto and A. J. Millis, *Theory of Mott insulator–band insulator heterostructures*, Phys. Rev. B, **70**, 075101 (2004)

[149] S. Okamoto and A. J. Millis, *Interface ordering and phase competition in a model Mott-insulator-band-insulator heterostructure*, Phys. Rev. B, **72**, 235108 (2005)

[150] Y. Okimoto, T. Katsufuji, Y. Okada, T. Arima, and Y. Tokura, *Optical spectra in $(La,Y)TiO_3$: Variation of Mott-Hubbard gap features with*

change of electron correlation and band filling, Phys. Rev. B, **51**, 9581 (1995)

[151] R. Otero, J. L. M. de Vidales, and C. de las Heras, *Synthesis and structural characterization of IMG thin films*, Journal of Physics: Condensed Matter, **10**(31), 6919 (1998)

[152] L. Paolasini, C. Vettier, F. de Bergevin, F. Yakhou, D. Mannix, A. Stunault, W. Neubeck, M. Altarelli, M. Fabrizio, P. A. Metcalf, and J. M. Honig, *Orbital Occupancy Order in V_2O_3: Resonant X-Ray Scattering Results*, Phys. Rev. Lett., **82**, 4719 (1999)

[153] J.-H. Park, L. H. Tjeng, A. Tanaka, J. W. Allen, C. T. Chen, P. Metcalf, J. M. Honig, F. M. F. de Groot, and G. A. Sawatzky, *Spin and orbital occupation and phase transitions in V_2O_3*, Phys. Rev. B, **61**(17), 11506 (2000)

[154] E. Pavarini, I. Dasgupta, T. Saha-Dasgupta, O. Jepsen, and O. K. Andersen, *Band-Structure Trend in Hole-Doped Cuprates and Correlation with $T_C^{max.}$*, Phys. Rev. Lett., **87**, 047003 (2001)

[155] A. Perucchi, C. Marini, M. Valentini, P. Postorino, R. Sopracase, P. Dore, P. Hansmann, O. Jepsen, G. Sangiovanni, A. Toschi, K. Held, D. Topwal, D. D. Sarma, and S. Lupi, *Pressure and alloying effects on the metal to insulator transition in $NiS_{2-x}Se_x$ studied by infrared spectroscopy*, Physical Review B (Condensed Matter and Materials Physics), **80**, 073101 (2009)

[156] W. E. Pickett, *Electronic structure of the high-temperature oxide superconductors*, Rev. Mod. Phys., **61**(2), 433 (1989)

[157] A. I. Poteryaev, J. M. Tomczak, S. Biermann, A. Georges, A. I. Lichtenstein, A. N. Rubtsov, T. Saha-Dasgupta, and O. K. Andersen, *En-*

hanced crystal-field splitting and orbital-selective coherence induced by strong correlations in V_2O_3, Physical Review B (Condensed Matter and Materials Physics), **76**, 085127 (2007)

[158] M. Potthoff and W. Nolting, *Weak-coupling approach to the semi-infinite Hubbard model: non-locality of the self-energy*, Zeitschrift für Physik B Condensed Matter, **104**(2), 265 (1997)

[159] M. Potthoff and W. Nolting, *Metallic surface of a Mott insulator–Mott insulating surface of a metal*, Phys. Rev. B, **60**(11), 7834 (1999)

[160] M. Potthoff and W. Nolting, *Surface metal-insulator transition in the Hubbard model*, Phys. Rev. B, **59**, 2549 (1999)

[161] T. Pruschke, D. L. Cox, and M. Jarrell, *Hubbard model at infinite dimensions: Thermodynamic and transport properties*, Phys. Rev. B, **47**(7), 3553 (1993)

[162] T. Pruschke, D. L. Cox, and M. Jarrell, *Transport Properties of the Infinite-Dimensional Hubbard Model*, EPL (Europhysics Letters), **21**(5), 593 (1993)

[163] S. N. Putilin, E. V. Antipov, O. Chmaissem, and M. Marezio, *Superconductivity at 94 K in $HgBa_2CuO_{4+\delta}$*, Nature, **362**(6417), 226 (1993)

[164] M. M. Qazilbash, M. Brehm, B. G. Chae, P.-C. Ho, G. O. Andreev, B. J. Kim, S. J. Yun, A. V. Balatsky, M. B. Maple, F. Keilmann, H. T. Kim, and D. N. Basov, *Mott Transition in VO_2 Revealed by Infrared Spectroscopy and Nano-Imaging*, Science, **318**, 1750 (2007)

[165] J. Rehr, *Theory and calculations of X-ray spectra: XAS, XES, XRS, and NRIXS*, Radiation Physics and Chemistry, **75**, 1547 (2006), Proceedings of the 20th International Conference on X-ray and Inner-Shell

Processes - 4-8 July 2005, Melbourne, Australia, Proceedings of the 20th International Conference on X-ray and Inner-Shell Processes

[166] T. M. Rice and W. F. Brinkman, *Effects of Impurities on the Metal-Insulator Transition*, Phys. Rev. B, **5**, 4350 (1972)

[167] W. R. Robinson, *High–temperature crystal chemistry of V_2O_3 and 1% chromium-doped V_2O_3*, Acta Crystallographica Section B, **31**, 1153 (1975)

[168] F. Rodolakis, P. Hansmann, J.-P. Rueff, A. Toschi, M. Haverkort, G. Sangiovanni, K. Held, M. Sikora, A. Congeduti, J.-P. Itie, F. Baudelet, P. Metcalf, and M. Marsi, *Electronic correlations in V_2O_3 studied with K-edge X-ray absorption spectroscopy*, Journal of Physics: Conference Series, **190**, 012092 (2009), 14th International Conference on X-Ray Absorption Fine Structure (XAFS14), 26-31 July 2009, Camerino, Italy

[169] F. Rodolakis, P. Hansmann, J.-P. Rueff, A. Toschi, M. Haverkort, G. Sangiovanni, A. Tanaka, T. Saha-Dasgupta, O. Andersen, K. Held, M. Sikora, I. Alliot, J.-P. Iti, F. Baudelet, P. Wzietek, P. Metcalf, and M. Marsi, *Inequivalent routes across the Mott transition in V2O3 explored by X-ray absorption*, Phys. Rev. Lett., **104**, 047401 (2010)

[170] F. Rodolakis, B. Mansart, E. Papalazarou, S. Gorovikov, P. Vilmercati, L. Petaccia, A. Goldoni, J. P. Rueff, S. Lupi, P. Metcalf, and M. Marsi, *Quasiparticles at the Mott Transition in V_2O_3: Wave Vector Dependence and Surface Attenuation*, Phys. Rev. Lett., **102**(6), 066805 (2009)

[171] M. J. Rozenberg, R. Chitra, and G. Kotliar, *Finite Temperature Mott Transition in the Hubbard Model in Infinite Dimensions*, Phys. Rev. Lett., **83**(17), 3498 (1999)

[172] M. J. Rozenberg, G. Kotliar, H. Kajueter, G. A. Thomas, D. H. Rapkine, J. M. Honig, and P. Metcalf, *Optical Conductivity in Mott-Hubbard Systems*, Phys. Rev. Lett., **75**, 105 (1995)

[173] M. J. Rozenberg, X. Y. Zhang, and G. Kotliar, *Mott-Hubbard transition in infinite dimensions*, Phys. Rev. Lett., **69**(8), 1236 (1992)

[174] A. N. Rubtsov, M. I. Katsnelson, and A. I. Lichtenstein, *Dual fermion approach to nonlocal correlations in the Hubbard model*, Phys. Rev. B, **77**, 033101 (2008)

[175] A. N. Rubtsov, V. V. Savkin, and A. I. Lichtenstein, *Continuous-time quantum Monte Carlo method for fermions*, Phys. Rev. B, **72**, 035122 (2005)

[176] A. Rüegg, S. Pilgram, and M. Sigrist, *Strongly renormalized quasi-two-dimensional electron gas in a heterostructure with correlation effects*, Phys. Rev. B, **75**, 195117 (2007)

[177] A. Sacchetti, E. Arcangeletti, A. Perucchi, L. Baldassarre, P. Postorino, S. Lupi, N. Ru, I. R. Fisher, and L. Degiorgi, *Pressure Dependence of the Charge-Density-Wave Gap in Rare-Earth Tritellurides*, Physical Review Letters, **98**, 026401 (2007)

[178] T. Saha-Dasgupta, O. K. Andersen, J. Nuss, A. I. Poteryaev, A. Georges, and A. I. Lichtenstein, *Electronic structure of V_2O_3: Wannier orbitals from LDA-NMTO calculations*, arXiv.org, **0**, 0907.2841 (2009)

[179] T. Saha-Dasgupta, J. Nuss, O. Jepsen, and O. K. Andersen (2009), Private communication

[180] O. Sakai and Y. Kuramoto, *Application of the numerical renormalization*

group method to the Hubbard model in infinite dimensions, Solid State Commun., **89**, 307 (1994)

[181] G. Sangiovanni, A. Toschi, E. Koch, K. Held, M. Capone, C. Castellani, O. Gunnarsson, S.-K. Mo, J. W. Allen, H.-D. Kim, A. Sekiyama, A. Yamasaki, S. Suga, and P. Metcalf, *Static versus dynamical mean-field theory of Mott antiferromagnets*, Phys. Rev. B, **73**, 205121 (2006)

[182] H. B. Schüttler and D. J. Scalapino, *Monte Carlo Studies of the Dynamics of Quantum Many-Body Systems*, Phys. Rev. Lett., **55**(11), 1204 (1985)

[183] K. Schwarz, P. Blaha, and G. K. H. Madsen, *Electronic structure calculations of solids using the WIEN2k package for material sciences*, Computer Physics Communications, **147**, 71 (2002)

[184] H. Schweitzer and G. Czycholl, *Resistivity and thermopower of heavy-fermion systems*, Phys. Rev. Lett., **67**(26), 3724 (1991)

[185] A. Sekiyama, H. Fujiwara, S. Imada, S. Suga, H. Eisaki, S. I. Uchida, K. Takegahara, H. Harima, Y. Saitoh, I. A. Nekrasov, G. Keller, D. E. Kondakov, A. V. Kozhevnikov, T. Pruschke, K. Held, D. Vollhardt, and V. I. Anisimov, *Mutual Experimental and Theoretical Validation of Bulk Photoemission Spectra of $Sr_{1-x}Ca_xVO_3$*, Phys. Rev. Lett., **93**(15), 156402 (2004)

[186] S. S. A. Seo, M. J. Han, G. W. J. Hassink, W. S. Choi, S. J. Moon, J. S. Kim, T. Susaki, Y. S. Lee, J. Yu, C. Bernhard, H. Y. Hwang, G. Rijnders, D. H. A. Blank, B. Keimer, and T. W. Noh, *Two-Dimensional Confinement of $3d^1$ Electrons in $LaTiO_3/LaAlO_3$ Multilayers*, Phys. Rev. Lett., **104**(3), 036401 (2010)

[187] J. Skilling, *Maximum Entropy and Bayesian Methods*, Kluwer Aca-

demic, Norwell (1989)

[188] J. C. Slater and G. F. Koster, *Simplified LCAO Method for the Periodic Potential Problem*, Phys. Rev., **94**, 1498 (1954)

[189] I. Solovyev, N. Hamada, and K. Terakura, t_{2g} *versus all 3d localization in LaMO$_3$ perovskites (M=Ti–Cu): First-principles study*, Phys. Rev. B, **53**(11), 7158 (1996)

[190] M. A. Subramanian, J. C. Calabrese, C. C. Torardi, J. Gopalakrishnan, T. R. Askew, R. B. Flippen, K. J. Morrissey, U. Chowdhry, and A. W. Sleight, *Crystal structure of the high-temperature superconductor $Tl_2Ba_2CaCu_2O_8$*, Nature, **332**(6163), 420 (1988)

[191] M. A. Subramanian et al., Physica (Amsterdam), **157 C**, 124 (1989)

[192] M. A. Subramanian et al., Physica (Amsterdam), **166 C**, 19 (1990)

[193] K. Takada, H. Sakurai, E. Takayama-Muromachi, F. Izumi, R. A. Dilanian, and T. Sasaki, *Superconductivity in two–dimensional CoO$_2$ layers*, Nature, **422**, 53 (2003)

[194] M. Takizawa, H. Wadati, K. Tanaka, M. Hashimoto, T. Yoshida, A. Fujimori, A. Chikamatsu, H. Kumigashira, M. Oshima, K. Shibuya, T. Mihara, T. Ohnishi, M. Lippmaa, M. Kawasaki, H. Koinuma, S. Okamoto, and A. J. Millis, *Photoemission from Buried Interfaces in SrTiO$_3$/LaTiO$_3$ Superlattices*, Phys. Rev. Lett., **97**, 057601 (2006)

[195] A. Tanaka and T. Jo, *Resonant 3d, 3p and 3s Photoemission in Transition Metal Oxides Predicted at 2p Threshold*, Journal of the Physical Society of Japan, **63**(7), 2788 (1994)

[196] T. Tetsuo Fujii, K. Tanaka, F. Marumo, and Y. Noda, *Structural behaviour of NiS_2 up to 54 kbar*, Mineralogical Journal, **13**, 448 (1987)

[197] B. T. Thole, P. Carra, F. Sette, and G. van der Laan, *X-ray circular dichroism as a probe of orbital magnetization*, Phys. Rev. Lett., **68**(12), 1943 (1992)

[198] T. Thole, *Memorial issue*, Journal of Electron Spectroscopy and Related Phenomena, **86**(1-3), 1 (1997)

[199] L. H. Tjeng, P. Rudolf, G. Meigs, F. Sette, C. T. Chen, and Y. U. Idzerda, In *Proc. SPIE*, vol. 1548, 160 (1991)

[200] J. G. Tobin, G. D. Waddill, and D. P. Pappas, *Giant x-ray absorption circular dichroism in magnetic ultrathin films of Fe/Cu(001)*, Phys. Rev. Lett., **68**(24), 3642 (1992)

[201] J. M. Tomczack, *Spectral and Optical Properties of Correlated Materials*, Ph.D. thesis, Ecole Polytechnique (2007)

[202] J. M. Tomczak and S. Biermann, *Optical properties of correlated materials: Generalized Peierls approach and its application to VO_2*, Phys. Rev. B, **80**, 085117 (2009)

[203] C. C. Torardi, M. A. Subramanian, J. C. Calabrese, J. Gopalakrishnan, E. M. McCarron, K. J. Morrissey, T. R. Askew, R. B. Flippen, U. Chowdhry, and A. W. Sleight, *Structures of the superconducting oxides $Tl_2Ba_2CuO_6$ and $Bi_2Sr_2CuO_6$*, Phys. Rev. B, **38**(1), 225 (1988)

[204] C. C. Torardi, M. A. Subramanian, J. C. Calabrese, J. Gopalakrishnan, K. J. Morrissey, T. R. Askew, R. B. Flippen, U. Chowdhry, and A. W. Sleight, *Crystal Structure of $Tl_2Ba_2Ca_2Cu_3O_{10}$, a 125 K Superconductor*, Science, **240**(4852), 631 (1988)

[205] A. Toschi, P. Hansmann, G. Sangiovanni, T. Saha-Dasgupta, O. K. Andersen, and K. Held, *Spectral properties of the Mott Hubbard insulator $(Cr_{0.011}V_{0.989})_2O_3$ calculated by LDA+DMFT*, Journal of Physics: Conference Series, **200**, 012208 (4pp) (2010)

[206] A. Toschi, A. A. Katanin, and K. Held, *Dynamical vertex approximation: A step beyond dynamical mean-field theory*, Phys. Rev. B, **75**(4), 045118 (2007)

[207] M. Uchida et al., *private communication* (2010), Tokyo

[208] G. van der Laan and B. T. Thole, *Strong magnetic x-ray dichroism in 2p absorption spectra of 3d transition-metal ions*, Phys. Rev. B, **43**(16), 13401 (1991)

[209] H. J. Vidberg and J. W. Serene, *Solving the Eliashberg equations by means of N-point Padé approximants*, Journal of Low Temperature Physics, **29**(3), 179 (1977)

[210] U. von Barth and L. Hedin, J. Phys. C: Solid State Phys., **5**, 1629 (1972)

[211] K. J. von Szczepanski, P. Horsch, W. Stephan, and M. Ziegler, *Single-particle excitations in a quantum antiferromagnet*, Phys. Rev. B, **41**(4), 2017 (1990)

[212] H. Wadati, Y. Hotta, A. Fujimori, T. Susaki, H. Y. Hwang, Y. Takata, K. Horiba, M. Matsunami, S. Shin, M. Yabashi, K. Tamasaku, Y. Nishino, and T. Ishikawa, *Hard x-ray photoemission study of $LaAlO_3/LaVO_3$ multilayers*, Phys. Rev. B, **77**, 045122 (2008)

[213] C. Weber, K. Haule, and G. Kotliar, *Optical weights and waterfalls in doped charge-transfer insulators: A local density approximation and*

dynamical mean-field theory study of $La_{2-x}Sr_xCuO_4$, Phys. Rev. B, **78**, 134519 (2008)

[214] T. Wegner, *Zur Theorie von Korrelations- und Temperatureffekten in Spektroskopien*, Ph.D. thesis, Humboldt–Universität zu Berlin (2000)

[215] P. Werner, A. Comanac, L. de' Medici, M. Troyer, and A. J. Millis, *Continuous-Time Solver for Quantum Impurity Models*, Phys. Rev. Lett., **97**, 076405 (2006)

[216] S. R. White, D. J. Scalapino, R. L. Sugar, E. Y. Loh, J. E. Gubernatis, and R. T. Scalettar, *Numerical study of the two-dimensional Hubbard model*, Phys. Rev. B, **40**, 506 (1989)

[217] F. Wooten, *Optical Properties of Solids*, Academic Press, New York (1972)

[218] X. Yao, Y.-K. Kuo, D. K. Powell, J. W. Brill, and J. M. Honig, *Magnetic susceptibility and heat-capacity studies of $NiS_{2-x}Se_x$ single crystals: A study of transitions at nonzero temperature*, Phys. Rev. B, **56**, 7129 (1997)

[219] J. Zaanen, O. Jepsen, O. Gunnarsson, A. Paxton, O. Andersen, and A. Svane, *What can be learned about high T_C from local density theory?*, Physica C: Superconductivity, **153-155**(Part 3), 1636 (1988), Proceedings of the International Conference on High Temperature Superconductors and Materials and Mechanisms of Superconductivity Part II

[220] J. Zaanen, G. A. Sawatzky, and J. W. Allen, *Band gaps and electronic structure of transition-metal compounds*, Phys. Rev. Lett., **55**, 418 (1985)

[221] H. W. Zandbergen *et al.*, Physica (Amsterdam), **159 C**, 81 (1989)

[222] F. C. Zhang and T. M. Rice, *Effective Hamiltonian for the superconducting Cu oxides*, Phys. Rev. B, **37**, 3759 (1988)

[223] V. Zlatic and B. Horvatic, *The local approximation for correlated systems on high dimensional lattices*, Solid State Communications, **75**(3), 263 (1990)

[224] E. Zurek, O. Jepsen, and O. K. Andersen, *Muffin-Tin Orbital Wannier-Like Functions for Insulators and Metals*, ChemPhysChem, **6**, 1934 (2005)

Acknowledgements

This thesis would not have been possible without the support of many people.

First of all, I thank Karsten Held for taking me in his group. I was really lucky to have been under his supervision, first in Stuttgart at the Max–Planck–Institute and then at the TU Vienna. The group he has created is exceptional and I am glad that I was a part of it from the very beginning.
Next, I would like to thank Alessandro Toschi. I think that it is, literally, impossible to care more about a student than he does. I did not only profit from the invaluable and innumerable discussions about physics with him, but he became also a close friend of mine. Besides physics, he introduced me to the Italian culture starting from the great food and coffee to the (really) crazy world of Italian politics.
Giorgio Sangiovanni and his family also became friends of mine and I thank them for all the great aperitivi and dinners we had here in Vienna and in Italy.

I started my studies in the department of Walter Metzner in the Max–Planck–Institute Stuttgart – I thank him and Dieter Vollhardt for the DMFT.
Ole Andersen and the people in his department were the most valuable collaborators for many projects presented in this thesis. Xiaoping Yang who did the NMTO calculations with great effort for all the nickelates provided the downfolded Hamiltonians. Maurits Haverkort provided great help and input on the XAS CI–calculations. Ove Jepsen, who gave substantial input on the NiS_2 systems, and Lilia Boeri were always there answering questions about

the DFT calculations.

Moreover, I would like to thank Giniyat Khaliullin and the experimentalists in the department of Bernhard Keimer whose work initiated the study on the nickelate heterostructures.

Ryotaro Arita in the University of Tokyo is our collaborator in Japan. He gave me the chance to discover Japan not only academically but also culturally – I really enjoyed my stays in Tokyo and am looking forward to my next visit. I would also like to thank Prof. Tokura and Prof. Nagaosa who initiated the collaboration on the bulk nickelates and Masaki Uchida who is doing the experimental work for these compounds.

I thank also all the experimentalists involved in the various projcts: Fanny Rodolakis in the group of Marino Marsi in Paris, Leonetta Baldassarre in Trieste and the group of Steffano Lupi in Rome.

Here at the TU Vienna I would like to thank all the members of the AG Held: Philipp Wissgott, Angelo Valli, Nico Parragh, Georg Rohringer, and Shiro Sakai. The relaxed and friendly atmosphere of our group is the best prerequisite for effective work and a high quality – something which cannot be valued enough. Also I would like to thank the CMS science college for financing my studies here in Vienna.

Zum Schluß möchte ich noch meinen Eltern danken, die mich immer ohne Kompromisse unterstützt haben. Auch Dir Sabrina danke ich dafür, daß Du immer für mich da bist.

Die VDM Verlagsservicegesellschaft sucht für wissenschaftliche Verlage abgeschlossene und herausragende

Dissertationen, Habilitationen, Diplomarbeiten, Master Theses, Magisterarbeiten usw.

für die kostenlose Publikation als Fachbuch.

Sie verfügen über eine Arbeit, die hohen inhaltlichen und formalen Ansprüchen genügt, und haben Interesse an einer honorarvergüteten Publikation?

Dann senden Sie bitte erste Informationen über sich und Ihre Arbeit per Email an *info@vdm-vsg.de*.

Sie erhalten kurzfristig unser Feedback!

VDM Verlagsservicegesellschaft mbH
Dudweiler Landstr. 99   Telefon +49 681 3720 174
D - 66123 Saarbrücken   Fax   +49 681 3720 1749
www.vdm-vsg.de

Die VDM Verlagsservicegesellschaft mbH vertritt

Printed by Books on Demand GmbH, Norderstedt / Germany